POTATO RADIO,
DIZZY DICE,

and More Wacky,
Weird, Experiments
from the
MAD SCIENTIST

POTATO RADIO, DIZZY DICE,

and More Wacky, Weird, Experiments from the MAD SCIENTIST

Joey Green

A Perigee Book

Warning: A responsible adult should supervise any young reader who conducts the experiments in this book to avoid potential dangers and injuries. The author has conducted every experiment in this book and has made every reasonable effort to ensure that the experiments are safe when conducted as instructed; however, neither the author nor the publisher assumes any liability for damages caused or injury sustained from conducting the projects in this book.

A Perigee Book
Published by The Berkley Publishing Group
A division of Penguin Group (USA) Inc.
375 Hudson Street
New York, New York 10014

Perigee trade paperback edition: July 2004

Visit our website at www.penguin.com

Library of Congress Cataloging-in-Publication Data

Green, Joey.
    Potato radio, dizzy dice, and more wacky, weird,
    experiments from the mad scientist / Joey Green.—
    1st Perigee ed.
        p. cm.
    Includes bibliographical references.
    ISBN 0-399-52992-6
    1. Science—Experiments. I. Title.

Q164 .G735 2004
793.8—dc22

                                        2003063946

Printed in the United States of America

            10 9 8 7 6 5 4 3 2 1

Contents

Introduction

I'm a Mad Scientist for one simple reason. I love making a mess. That's how this book came to be. You see, I'm always looking for exciting ways to make a bigger mess. I realized that one of the easiest ways to make a mess is to throw confetti. Now I don't know about you, but throwing a handful of confetti just doesn't make a big enough mess for me. Have you ever seen huge clouds of confetti showering down on a crowd of people during a parade, a concert, or big celebration? Now we're talking.

But how could I ever do that? Well, the simple way to shower a roomful of people with confetti is to hire a company to fire off a confetti blaster. A confetti blaster is basically a cannon powered by a canister of compressed air or carbon dioxide. Unfortunately, hiring a company to fire off a confetti cannon costs a lot of money. I could always buy my own confetti cannon, but I wanted a way to blast confetti without spending a million dollars. Then I realized, "Hey, wait a minute, I'm a Mad Scientist! I can make my own confetti blaster!" And so, I locked myself in the garage and tried to figure out how I could use a bunch of old sprinkler pipes, wires, switches, a bicycle pump, and some old-fashioned American ingenuity to build a homemade confetti blaster.

After designing my kooky contraption, I needed confetti. Of course, I could take a pair of scissors and spend the next three days cutting pieces of tissue paper into tiny little squares. I figured it would be a lot easier to go to a store and buy a bag of confetti. But where do you buy confetti that fits into a confetti blaster? A quick search of the Internet revealed that several companies sell a wide variety of confetti especially made for confetti cannons. Unfortunately, I'm not the kind of guy who likes to spend seven dollars for a plastic bag filled with chopped up pieces of paper.

So I looked around the garage for something else that might make a good substitute for confetti. I spotted a box of foam peanuts. But would foam peanuts work in a confetti blaster? There was only one way to find out. I broke the foam peanuts into small pieces, filled the blaster with them,

pumped the bicycle pump to fill the blaster with compressed air, and pushed the button. The confetti blaster fired off a beautiful shower of foam confetti in the air, much to my delight. Of course, anytime I want real confetti, I can spend three days cutting pieces of tissue paper into small squares. That's the beauty of being a Mad Scientist.

And that's how this book came to be. You see, I had so much fun making the confetti blaster that I just had to write another book of mad science experiments to share with my fellow Americans—so now you too can experience the joy of firing off a shower of foam peanuts in the privacy of your own backyard. But I didn't stop there. I also came up with a cool way to make blacklight ink using Tide, devised an awesome way to make a latex glove inflate itself, and figured out how you can make a working submarine out of an empty soda bottle, a balloon, and a handful of nickels. I think this is my most exciting *Mad Scientist Handbook* yet. Get ready to have a blast.

Ant Farm

WHAT YOU NEED

- ☐ Ruler
- ☐ Pencil
- ☐ Two 8-inch lengths of ½-inch-square wood
- ☐ Electric drill with ⅛-inch bit
- ☐ Four pieces of wire screening (½-inch square)
- ☐ Epoxy glue
- ☐ Three 11-inch lengths of ½-inch-square wood
- ☐ Two sheets of glass (9 by 11 inches) from two photograph frames
- ☐ 10-inch length of ½-inch-square wood

- ☐ Elmer's Carpenter's Wood Glue
- ☐ 12-inch length of ½-inch-square wood
- ☐ Sandpaper
- ☐ 5-by-13-inch piece of plywood
- ☐ Spoon
- ☐ Clean, empty mayonnaise jar with lid
- ☐ White posterboard
- ☐ Digging spade
- ☐ Q-tips cotton swab
- ☐ Honey
- ☐ Piece of black fabric (12 by 20 inches)
- ☐ Cereal crumbs, dead insects, finely chopped vegetables or fruits

WHAT TO DO

Measure 1 inch from one end of one of the 8-inch lengths of ½-inch-square wood and make a mark. With adult supervision, drill a ⅛-inch-diameter hole in the center of the mark. Repeat with the second 8-inch length of wood. Place a ½-inch-square piece of screening over both sides of each hole and adhere in place with epoxy glue. Let dry.

Place one of the 11-inch lengths of wood along one of the 11-inch sides of one of the panes of glass. Place the 8-inch lengths along the 9-inch sides, with their holes farthest from the 11-inch wood and facing each other. Place the 10-inch length of wood between the two open ends of the 8-inch lengths to make certain they fit snugly. Using epoxy glue, adhere the 11-inch piece of wood and two 8-inch pieces of wood to the glass. Make certain the epoxy glue forms a continuous seal between the glass and wood. Let dry for one hour.

Apply a solid line of epoxy glue along the edge of the two wooden sides and one bottom. Place the second piece of glass over the glued edges to fit over the frame. Let dry overnight.

Using the wood glue, adhere the 10-inch length of wood to the 12-inch length of wood, centering it to allow 1 inch clearance at both ends. Let dry overnight.

Make certain the 12-inch lid fits snugly into the opening at the top of the frame. If the lid is too large, use sandpaper to gently shave the wood lid.

Measure a horizontal line dividing the 5-by-13-inch piece of plywood in half along the length of the wood. Using wood glue, adhere the bottom of the frame to the line, leaving 1 inch free at both ends. Adhere the remaining two 11-inch lengths of wood along both sides of the base of the frame. Let dry overnight.

Locate an anthill outdoors (easily found in a dry, sandy field), and using a spoon, fill a jar with about one hundred ants and soil. Place the white posterboard flat on the ground, and using the small spade, dig deeper, breaking up clumps of soil on the posterboard until you locate the queen—an ant that is significantly larger and paler.

Set the ant farm on the posterboard and fill the frame to 1 inch below the

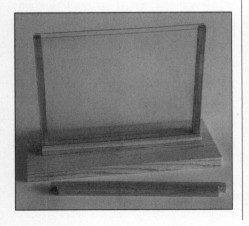

airholes with soil and ants. Dip one end of a Q-tips cotton swab in honey and stick it in the soil. Place the lid securely on the ant farm.

Cover the ant farm for twenty-four hours by draping the piece of black fabric over the frame.

Use an eyedropper to moisten the soil and feed the ants honey on a Q-tip cotton swab once a week. You can also feed the ants cereal crumbs, dead insects, and finely chopped vegetables or fruits. When not viewing the ant farm, cover the frame with the black fabric.

WHAT HAPPENS
The ants tunnel a complex network of tunnels and chambers that can be seen through the glass.

WHY IT WORKS
Ants build nests composed of tunnels and chambers in soil. The thin wall of soil sandwiched between two plates of glass gives the ants limited space to build their nest, giving you a cross-section view of their activity. Covering the ant farm with the black fabric causes the ants to work faster because ants work best in the dark.

BIZARRE FACTS
- An ant farm is called a formicarium.
- Some ants build nests that extend up to forty feet beneath the ground.
- Ants lay a scent trail of pheromones to lead other ants from the nest to a food source. Ants release other pheromones with a distinctive smell to warn nestmates of danger.
- Ants use their antennae to sense hearing, smell, touch, and taste.
- Ant nests generally include a chamber to house the queen and her eggs, several chambers to serve as nurseries filled with growing young, and chambers for food storage.
- Ants have lived on earth for more than 100 million years, according to dated fossils preserved in amber.
- All worker ants—the vast majority of any ant colony—are female.
- A queen ant lives anywhere from ten to twenty years, laying thousands of eggs during her lifetime. Worker ants live from less than one year to more than five years. Male ants live less than a few months.
- Ants can lift between ten and fifty times their weight.
- The heart of an ant is a long tube that stretches from the brain to the end of the ant's body.
- Male and queen ants have wings, but only use them once during the mating flight, after which the males die and the queens tear off their own wings.
- Ants eat more than half the termites hatched each year in the tropics.
- In the 1954 science-fiction movie *Them!*, nuclear tests in the desert generate gigantic mutant ants that menace cities in the American Southwest. This Cold War–era film stars James Whitmore, Edmund

Gwenn, James Arness, Joan Weldon, and Fess Parker.

- Fear of ants is called myrmecophobia.
- *The Atom Ant Show*, created by Hanna-Barbera and broadcast on NBC from 1965 to 1967, featured the cartoon adventures of superhero Atom Ant, an animated ant who could fly, lift ten times his weight, and communicate with his superiors with his antennae.
- Rock singer Adam Ant, born Stuart Leslie Goddard, changed his name to Adam after the biblical story of the Garden of Eden and named his band the Ants, choosing an insect name along the lines of the Beatles.

- The black bulldog ant, indigenous to Australia and Tasmania, can kill a human being.
- In the 1998 animated movie *Antz*, Woody Allen provides the voice of Z, a neurotic worker-ant who attempts to assert his individuality and find his true self within the confines of an oppressively conformist ant colony.

Mmmmm, Mmmmm, Puke!

Ant nestmates share food through mouth-to-mouth regurgitation.

Anti-Gravity Bucket

WHAT YOU NEED

- ☐ Plastic bucket with handle
- ☐ Clothesline rope (6 feet in length)
- ☐ 2 quarts water
- ☐ Work gloves

WHAT TO DO

Tie one end of the rope securely to the handle of the bucket (with several knots). Pour the water into the bucket. Wearing the gloves (to avoid getting rope burn), wrap the free end of the rope around one hand. Standing outdoors, swing the bucket from the rope like a lasso, making the bucket circle around you. Continue swinging the bucket around so it gathers enough momentum to circle around you parallel to the ground. Then tip the angle of the rope so the bucket circles at an angle over your head.

WHAT HAPPENS

The water remains inside the circling bucket.

WHY IT WORKS

Centrifugal force—the force created by whirling the rope—causes the water to move toward the bottom of the bucket. Centripetal force, however, holds the water in the bucket. The rope applies centripetal force to the whirling bucket, pulling it inward and preventing the bucket (and the water) from moving in a straight line.

BIZARRE FACTS

- The faster you twirl the bucket around you, the stronger the pull (or centrifugal force) on the rope.
- Most people confuse centrifugal force with centripetal force. Centrifugal force is the outward force on the object rotating about an axis. Centripetal force is the inward force on that same object. The forces acting on the same object are equal and opposite, in keeping with Newton's third law of motion ("For every action there is an equal and opposite reaction").
- When you ride on a merry-go-round, your body wants to shoot off in a straight line, but is held back by centripetal force (your grip on the horse).
- The earth's gravity exerts a centripetal force on the moon and artificial satellites, preventing them from shooting out into space and keeping them orbiting the planet. Similarly, the sun's gravity exerts a centripetal force on the earth, preventing our planet from shooting out into space.

- A spinning object, such as a top, stays in motion due to angular momentum.
- To perform a spectacular spin, an ice skater starts spinning with his arms outstretched, creating a larger spin diameter and greater angular momentum. By suddenly pulling in his arms, the skater reduces his diameter, causing his body to spin faster to conserve angular momentum.
- The phrase "drop in the bucket" means an unsubstantial contribution, and the phrase "kick the bucket" means to die.
- In the novel *Charlie and the Chocolate Factory* (and the movie *Willy Wonka and the Chocolate Factory*), Charlie's last name is Bucket.

The Earth Is Not a Sphere

The earth, flattened at the poles and bulging at the equator, is actually an oblate spheroid. In other words, the earth's equatorial circumference (24,901.55 miles) is greater than its polar circumference (24,859.82 miles). According to Sir Isaac Newton, this bulge is caused by the rotating earth's centrifugal force.

Basket Case

WHAT YOU NEED

☐ Disposable, reusable plastic storage container (approximately 6 inches square)
☐ Electric drill with 1-inch bit
☐ Large pot or bathtub full of water
☐ Wax paper (4 inches square)
☐ Paper towel (4 inches square)

WHAT TO DO

With adult supervision, drill six or more holes in the bottom of the plastic storage container.

Gently place the perforated basket on the surface of the water so that it floats. Place the square of wax paper into the basket and observe. Carefully remove the wax paper from the basket. Place the square of paper towel into the basket and observe.

WHAT HAPPENS

The plastic basket floats on the surface of the water. When you place the square of wax paper in the basket, the basket continues to float. When you place the square of paper towel in the basket, the basket sinks.

WHY IT WORKS

Surface tension is a force on the surface of a liquid that causes the surface of the liquid to behave as if a thin,

elastic skin covers it. The plastic basket, while perforated, deforms the surface tension skin on the water without breaking the surface tension skin. Adding the sheet of wax paper to the basket does not break the surface tension skin of the water. The paper towel, however, absorbs water, breaking the surface tension and making the basket heavier than the water, causing the basket to sink.

BIZARRE FACTS

- Surface tension is caused by cohesion, a force that causes the molecules of the same substance to be attracted to each other.
- Surface tension causes raindrops and dewdrops to form as spheres. The water molecules, pulling in all directions, are attracted to each other. The outer water molecules, however, are only pulled inward—having no water molecules beyond the surface to pull them outward.

This inward attraction pulls the water molecules together to form a sphere—a shape with the minimum surface area.

- Adding a few drops of vegetable oil to a pot of water breaks the surface tension of the water, preventing spaghetti from sticking together in a boiling pot of water.
- Surface tension prevents water from washing dirt from skin or grease from plates. Soap, dissolved in water, reduces the surface tension of the water by separating the water molecules, allowing the water to wash the dirt and grease away.
- Surface tension is caused by the inward pull on molecules at the surface of a liquid. A molecule of a liquid that is below the surface is pulled in every direction by the surrounding molecules. A molecule of a liquid that is on the surface is pulled only downward and to the sides by the surrounding molecules, causing surface tension.

The Floating Paper Clip

You can fill a glass with water and carefully place a steel paper clip on the surface of the water so that the surface tension allows the paper clip to float—the same way insects walk on the surface of a still pond.

Blacklight Ink

WHAT YOU NEED

☐ Liquid Tide
☐ Paper cup
☐ Paintbrush
☐ White posterboard
☐ Blacklight

WHAT TO DO

Fill a paper cup halfway with Liquid Tide. With the paintbrush, paint designs or a message on the white posterboard. Let sit until the liquid detergent dries, turning invisible. In a dark room, turn on the blacklight and hold the poster near the light.

WHAT HAPPENS

The liquid detergent painted on the white posterboard appears invisible in ordinary light, but glows purplish white under the blacklight.

WHY IT WORKS

Liquid Tide contains a fluorescent chemical that is activated by the ultraviolet rays produced by a blacklight.

The fluorescent chemical in the Liquid Tide converts the ultraviolet light into visible light.

BIZARRE FACTS

- You can also make bubbles with Liquid Tide that glow under a blacklight. Simply dilute the Liquid Tide with enough water to make bubbles.
- Murine Tears eyedrops glow under a blacklight as fluorescent yellow.
- Cat urine glows under a blacklight.
- Ultraviolet rays (also known as black light) are a form of light invisible to the human eye and lie beyond the violet end of the visible spectrum.
- *Consumer Reports* claims, "No laundry detergent will completely remove all common stains," and reports very little difference in performance between major name-brand powdered detergents.

- Every year, researchers for Procter & Gamble duplicate the mineral content of water from all parts of the United States and wash fifty thousand loads of laundry to test Tide detergent's consistency and performance.
- Ultraviolet rays can cause sunburn and penetrate clouds, which is why a person can get sunburned on an overcast day.
- Lightning produces ultraviolet rays.
- A blacklight lamp produces ultraviolet rays when an electric current passes through the glass tube filled with a gas or vapor.
- Ultraviolet rays with wavelengths shorter than 320 nanometers can be used to kill bacteria and viruses.
- Ultraviolet rays from the sun produce vitamin D in the human body.

Blacklight Bubbles

For a totally psychedelic experience, mix one-half cup Tide with one-half cup water, and use the solution to blow bubbles under the blacklight. The bubbles will glow purplish white.

Bottle Rocket

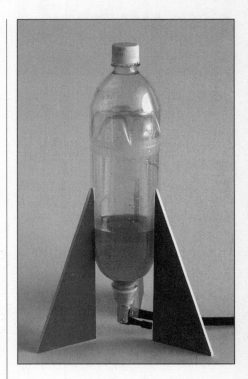

WHAT YOU NEED

☐ Electric drill with $\frac{1}{32}$-inch bit

☐ Cork

☐ Scissors

☐ Clean, used Playtex Living glove

☐ Needle adapter for inflating balls

☐ Ruler

☐ Pencil

☐ Foam-core board

☐ X-acto knife or single-edge razor blade

☐ Two 1-liter plastic soda bottles

☐ Clear packaging tape

☐ 1 cup water

☐ Funnel

☐ Bicycle pump and connector

WHAT TO DO

With adult supervision, drill a hole through the center of the cork. Using scissors, cut a section from a finger of the rubber glove to fit tightly around the cork as a sleeve. Insert the needle adapter through the hole in the cork so it fits snugly.

On the foam-core board, measure a right triangle with a base 3 inches long and a height of 8 inches long. With adult supervision, carefully cut out the triangle with the X-acto knife or single-edge razor blade. Use this triangle as a template to create two more triangles.

Turn one of the plastic bottles upside down and attach the triangular fins at equal distances around the bottle with strips of clear packaging tape, so that the fins support the rocket.

With adult supervision, use the X-acto knife or single-edge razor blade to cut the top 4 inches from the second plastic soda bottle. Place the top over the bottom of the first plastic soda bottle to create the rocket cone, and secure in place with clear packaging

tape. Secure the bottle cap in place on the top of the rocket.

Using the funnel, hold the rocket upside down and fill the bottle with 1 cup water. Insert the cork into the neck of the bottle so it fits tightly. Connect the foot pump to the needle adapter.

Set the rocket on its fins on a flat surface outdoors—away from trees, telephone wires, electrical cables, and buildings. With adult supervision and standing a safe distance away from the rocket, pump air into the bottle using the bicycle pump.

WHAT HAPPENS
The cork eventually pops from the neck of the bottle, and water and air shoot out, sending the rocket high into the air.

WHY IT WORKS
The air pressure builds up inside the bottle until the cork pops out, allowing the water and air to escape from the bottle. Consequently, the bottle rocket is propelled upward because, as Sir

Isaac Newton stated in his third law of motion, for every action there is an opposite and equal reaction.

BIZARRE FACTS
■ The Space Shuttle and *Saturn V* rockets are propelled into space due to Newton's third law of motion. When fuel burns inside the rocket's combustion chamber, the resulting hot gases shoot from the nozzle. The reacting force sends the rocket upward.

■ Sir Isaac Newton, an ordained priest in the Church of England, taught mathematics at Cambridge University.

■ Newton invented calculus.

■ Newton's third law of motion explains why guns kick back violently when fired.

■ When you shoot a cue ball into a stationary ball on a pool table, both balls push each other with equal force in opposite directions. The stationary ball gains momentum and the cue ball loses the exact same amount of momentum.

Making a Splash

You can demonstrate Newton's third law by simply jumping off an inflatable raft in a swimming pool. When you jump forward, the raft moves backward with equal momentum.

Colored Flames

WHAT YOU NEED

- ☐ 1 tablespoon 20 Mule Team borax (available at the supermarket)
- ☐ 1 tablespoon salt
- ☐ 1 tablespoon copper sulfate (available as Gordon's Bordeaux Mixture at the hardware store or garden center)
- ☐ 1 tablespoon boric acid (available at the hardware store or drugstore)
- ☐ 1 tablespoon Morton's Salt Substitute or No-Salt (available at the supermarket)
- ☐ 1 tablespoon calcium chloride (available as Leslie's Hardness Plus at Leslie's pool supply store)
- ☐ Muffin tin
- ☐ Six cotton balls
- ☐ Rubbing alcohol (70 percent isopropyl)
- ☐ Box of baking soda (or pitcher of water)
- ☐ Matches

WHAT TO DO

Place the tablespoon of borax in one of the empty compartments of the muffin tin.

Saturate a cotton ball with alcohol, then dip the cotton ball in the borax until it is coated with the powder. Let the cotton sit in the compartment.

Secure the cap on the alcohol bottle and move it to the opposite side of the room. Wash your hands with soap and water to remove any alcohol or chemical residue. Move the muffin tin outdoors. Open the box of baking soda,

and keep it (or a pitcher of water) nearby for use as a fire extinguisher. (Sprinkling baking soda over a fire will extinguish the fire.)

With adult supervision, carefully use the matches to ignite the cotton ball. Observe. Extinguish the flame.

Repeat this process until you have dipped all the cotton balls in a differ-ent chemical powder, placed each powdered cotton ball in a separate compartment in the muffin tin, and lit it on fire.

WHAT HAPPENS

The borax burns green. Salt is sodium chloride, which burns yellow. The copper sulfate burns with a green flame. Boric acid burns blue-green flames. No-Salt contains potassium chloride, which gives the flame a violet color. Calcium chloride burns orange-red.

WHY IT WORKS

In 1777, French chemist Antoine Lavoisier proved that fire is the heat and light that results from the rapid union (or combustion) of oxygen with other substances. The color of the flame depends on the substance being

You're Fired

During World War II, the Pentagon developed a one-ounce incendiary bomb that could be strapped to the chest of a bat and then dropped over Japanese cities, where the bat would then chew through the straps, detonating the bomb. The bomb would flare for eight minutes with a twenty-two-inch flame. The Pentagon planned to use these bat bombs to set fire to Japan's wood houses and buildings. After researching "Project X-ray" for two years and re-cruiting two million bats from the American Southwest, the Army tested the bat bombs in New Mexico. During the testing, several bats escaped, setting fire to a large aircraft hangar and a general's staff car. The Navy took over the project and decided to freeze the bats into hibernation before dropping them out of the bombers. In a test run in August 1944, the frozen bats were dropped out of the bombers, remained asleep, and penetrated into the earth. Project X-ray was immediately suspended, having cost two million dollars.

united with the oxygen. In this experiment, the alcohol provides the fuel and the cotton provides the kindling to ignite the other chemicals.

BIZARRE FACTS

- When oxygen unites with another substance, either slowly or rapidly, the process is called oxidation.
- Oxygen unites slowly with iron to cause rust, rather than combustion.
- When oxygen unites with gunpowder or dynamite, the rapid oxidation creates huge volumes of gas that expand rapidly and violently, producing an explosion.
- Ever since the discovery of fire, people have tried to figure out how to harness the energy of fire to create more light. They discovered that wood dipped in pitch would burn brighter and longer. Eventually, people discovered that a wick placed in a bowl of oil would burn brightly.
- The lit wick of a candle melts the surrounding wax. The porous wick absorbs the molten wax, which travels up the wick, feeding the flame with fuel.
- Scientists use flame tests to identify the elements present in a chemical. The chemists grind up the chemical sample, mix it with methylated spirit, and light it with a match. The color of the flame helps indicate what type of salts are in the chemical.
- The phrase "to get fired" or "to fire someone" originated from the ancient practice of setting fire to the home of an unwanted member of a tribe to rid him from a village or clan.
- The cap on a fire hydrant is called the bonnet.

Confetti Blaster

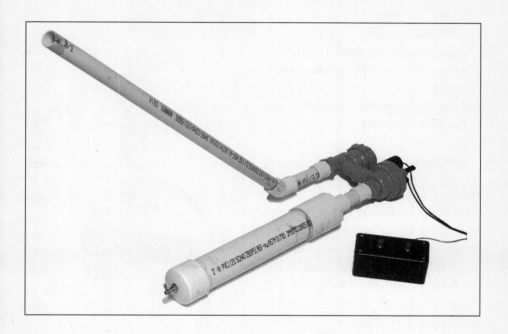

WHAT YOU NEED

- Drill with ⅜-inch bit and ¼-inch bit
- 2-inch-diameter PVC end cap
- Metal automotive tire valve with nuts to tighten (available at an automotive supply store)
- Adjustable wrench
- Utility knife
- Teflon tape
- 24-volt plastic sprinkler valve with ¾-inch-diameter threaded male pipe connections
- Two PVC pipe adapters from ¾-inch to ¾-inch female screw head
- 2-foot-long 2-inch-diameter PVC pipe
- 2-inch-diameter PVC coupler
- PVC pipe reducer fitting from 2-inch diameter to ¾-inch diameter
- 3-inch-long ¾-inch-diameter PVC pipe
- 3-foot-long ¾-inch-diameter PVC pipe
- ¾-inch 45-degree-angle PVC coupler
- 6-inch-long ¾-inch-diameter PVC pipe
- PVC pipe primer
- PVC pipe glue
- Wire cutters
- 15 feet of two-conductor 24-gauge wire
- Electrical tape
- Plastic project enclosure box (5 by 2½ by 2 inches)
- SPST on/off toggle switch

- ☐ SPST momentary push-button switch
- ☐ Two 9-volt battery snap connectors
- ☐ 6-inch length of single-conductor 24-gauge wire
- ☐ Solder gun
- ☐ Solder
- ☐ Two 9-volt batteries
- ☐ Phillips head screwdriver
- ☐ Bicycle pump with air pressure gauge
- ☐ Tissue
- ☐ 4-foot wooden dowel (¼-inch diameter)
- ☐ Box of foam peanuts
- ☐ Wax paper
- ☐ Masking tape

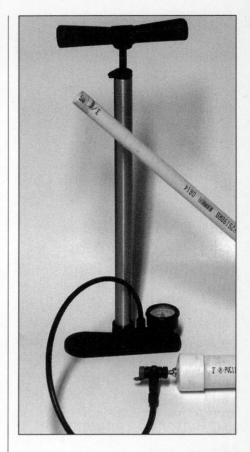

WHAT TO DO

With adult supervision, drill a ⅜-inch hole in the center of the end cap. Unscrew the tightening nut from the metal automotive tire valve, insert the tire valve through the inside of the end cap (making sure the rubber gasket covers the drilled hole), and tighten the nut securely in place with the adjustable wrench.

Using the utility knife, carefully scrape off any burrs left from the ends of the cut pieces of PVC pipe to assure a clean seal when the pipes are securely sealed together airtight.

Wrap a piece of Teflon tape around each one of the threaded male joints on the 24-volt sprinkler valve. (Do not use the PVC pipe glue on any threaded joints.) Screw a PVC pipe adapter (with the threaded female joint) onto each one of the threaded male joints of the sprinkler valve. Secure tightly.

Coat the outsides of the ends of each PVC pipe and the inside of each PVC coupler, the outside and inside of the PVC pipe reducer fitting, and the end cap with PVC pipe primer. Using the PVC pipe glue, attach the 2-inch-diameter PVC end cap (previously prepared with the tire valve) to the end of the 2-foot-long 2-inch-diameter PVC pipe. To the other end of the 2-foot-long 2-inch-diameter PVC pipe, glue the 2-inch-diameter PVC coupler. Glue the PVC pipe reducer fitting inside the 2-inch-diameter PVC coupler. Glue one end of the 3-inch-long ¾-inch-diameter PVC pipe into the ¾-inch hole of the PVC pipe reducer fitting. Glue the open end of the 3-inch-long PVC pipe into the adapter at the bottom of the 24-volt sprinkler valve beneath the chamber with the wires (and the top secured with eight screws). This completes the air-compression chamber.

Glue one end of the 3-foot-long ¾-inch-diameter PVC pipe into the ¾-inch 45-degree-angle PVC coupler. Glue one end of the 6-inch-long ¾-inch-diameter PVC pipe into the open end of the 45-degree-angle PVC coupler. Glue the open end of the 6-inch-long ¾-inch-diameter PVC pipe into the adapter at the bottom of the 24-volt sprinkler valve beneath the chamber without the wires. Make certain that when the device sits on the floor, the 3-foot-long pipe rises from the floor at a 45-degree angle. This completes the barrel of the confetti blaster. Allow the PVC pipe glue to dry for twenty-four hours before using the blaster.

Use the wire cutters to strip ½ inch of plastic coating off the four ends of the two-conductor wires. Wire one end of each wire to the ends of the two wires coming from the 24-volt sprinkler valve. Individually wrap each connection with electrical tape to prevent the connections from contacting each other. Then tape the two wrapped connections together.

Drill two ¼-inch holes 2 inches apart in the top of the plastic project enclosure box. Insert the on/off toggle switch into one hole and secure in place with the nut. Insert the momentary push-button switch into the other hole and secure in place with the nut.

Drill a hole in the center of one of the 5-inch-long sides of the plastic box itself. Tie a knot in the free end of the 15-foot wire, 6 inches from the two free ends. Insert the two free ends into the hole you drilled in the side of the plastic box and tie another knot in the double wire inside the box to secure the wire in place.

Wire one of the two free wires to one wire from a 9-volt battery snap connector. Wire the free wire from the 9-volt battery snap connector to one of the wires from the second 9-volt battery snap connector. Wire the remaining free wire from the second 9-volt battery snap connector to one pole on the on/off toggle switch.

Use the wire cutters to strip ½ inch of plastic coating off the two ends of the 6-inch-long single-conductor wire. Wire one end of the 6-inch-long wire to the free pole on the on/off toggle switch. Wire the remaining end of the 6-inch-long wire to a pole on the momentary push-button switch. Wire the remaining free end of the 15-foot wire to the free pole on the momentary push-button switch.

Solder the connections. When the solder cools, wrap the connections with electrical tape (to prevent the batteries that will be inside the box from accidentally bridging the contacts). Snap the batteries into place. (The two 9-volt batteries provide a total of 18 volts, which, surprisingly, is sufficient to activate the 24-volt sprinkler valve.) Using a Phillips head screwdriver, screw the cover panel into place on the project enclosure box.

After allowing the PVC pipe glue to dry for twenty-four hours before using the blaster, attach the bicycle pump to the tire valve. Pump air into the chamber until you achieve 80 pounds per square inch (psi). Fill a bathtub with 6 inches of water, and hold the device underwater to check for leaks. If you encounter a leak, use an aerosol can of Fix-a-Flat (available at automotive stores) to fix the leak.

Ball up a tissue, insert it into the open end of the ¾-inch pipe, and use the dowel to push the tissue down the tube just before the first joint. Break the foam peanuts into small pieces, and pour 1 cup of them into the open end of the ¾-inch pipe.

Cover the open end of the barrel with a piece of wax paper and secure it in place with masking tape around the circumference (not over the top of the wax paper).

Attach the bicycle pump to the tire valve and pump air into the chamber until you achieve 80 psi. (A 24-volt sprinkler valve can generally handle up to 150 psi.)

Position the confetti blaster on the ground outside, aiming the open end of the ¾-inch pipe away from people or pets and toward an open area.

Flip the on/off switch to on. Press the momentary push-button switch.

WHAT HAPPENS

The foam peanuts blast through the wax paper, causing a popping sound, and shoot up to forty feet into the air.

WHY IT WORKS

When you press the button, the batteries activate the electromagnet inside the sprinkler valve, opening the valve and allowing the air pressure stored in the first chamber to rush out the barrel of the air gun, forcing out the contents of the barrel. The homemade air cannon can also shoot off streamers or confetti. To obtain streamers or confetti, visit www.artistryinmotion.com. You will need confetti that fits into a ¾-inch cannon. To shoot larger loads of confetti, use 1-inch-diameter PVC pipe for the cannon.

BIZARRE FACTS

- Professional confetti cannons commonly use canisters of compressed carbon dioxide to power the blasts. Canisters of carbon dioxide can be quite costly.
- In 1839, German pharmacist Eduard Simon isolated a strong, versatile substance from natural resin, without knowing what he had discovered. Nearly a century later, German organic chemist Hermann Staudinger identified Simon's discovery, comprised of long chains of styrene molecules, as a plastic polymer—polystyrene.
- In 1930, scientists at Badische Anilin & Soda-Fabrik (BASF), a German company founded in 1861, commercially manufactured polystyrene from erethylene and benzine.
- In 1937, Dow Chemical Company introduced polystyrene in the United States.
- In the 1940s, Dow Chemical scientist Ray McIntire, working to develop a new flexible electrical insulator, tried to create a new rubber-like polymer by combining styrene with isobutylene, a volatile liquid, under pressure. Instead of producing an elastic polymer, his experiment accidentally yielded foam polystyrene, filled with air bubbles and thirty times lighter than regular polystyrene.
- In 1954, the Dow Chemical Company introduced its polystyrene foam product under the brand name Styrofoam. The company first introduced Styrofoam as a flotation material in life rafts and lifeboats, then as a thermal insulation material for cold store floors, wall and ceiling panels, and pipes.
- Polystyrene foam products are 95 percent air and only five percent polystyrene.
- The fully closed cell structure of foam polystyrene makes it highly resistant to water absorption and an excellent insulation material.
- Foamed polystyrene is used to make cups, bowls, plates, trays, clamshell containers, meat trays, egg cartons, and protective packaging for shipping electronics and other fragile items.
- Solid polystyrene is used to make cutlery, yogurt and cottage cheese containers, cups, clear salad bar containers, and video and audiocassette housings.
- Polystyrene foam products are produced primarily using two types of blowing agents: pentane and carbon dioxide.

Disappearing Peanuts

WHAT YOU NEED

- ☐ Rubber gloves
- ☐ Protective goggles
- ☐ Protective breathing mask
- ☐ 1-quart can of acetone (available at the hardware store)
- ☐ Galvanized metal bucket (not plastic)
- ☐ Large box of polystyrene peanuts

WHAT TO DO

With adult supervision and wearing the rubber gloves and protective goggles and breathing mask, carefully pour the acetone into the bucket. Pour the polystyrene peanuts from the box into the bucket, gently swirling the bucket to swish the acetone around.

WHAT HAPPENS

The polystyrene peanuts vanish and only a rubbery goo remains floating in the acetone.

WHY IT WORKS

The acetone acts as a solvent, dissolving the links between the molecules in the polystyrene peanuts and releasing the trapped gas in the foam, leaving behind the liquid hydrocarbon styrene used to create the peanuts.

BIZARRE FACTS

- Polystyrene is a solid plastic that results from the homopolymerization of the liquid hydrocarbon styrene.
- Peanuts are actually considered a legume, not a nut.
- Polystyrene foam—used to make disposable coffee cups, fast-food hamburger boxes, and packing peanuts—is considered environmentally hazardous because it takes up space in landfills, requires decades to decompose, and its manufacture causes the release of hazardous chemicals.
- In 1990, McDonald's phased out foam packaging from its restaurants in the United States.
- Polystyrene foam can be recycled by turning the plastic into pea-size pellets for use in wall insulation and industrial packaging.
- The comic strip "Peanuts," created by Charles Schulz and featuring Charlie Brown and Snoopy, was named "Peanuts" (meaning "tiny people") by King Features Syndicate against Schulz's wishes.

Working for Peanuts

Mr. Peanut was designed in 1916 by a Suffolk, Virginia, schoolchild who won five dollars in a contest sponsored by Planters Peanuts.

Dizzy Dice

WHAT YOU NEED

- ☐ Three pieces of white foam-core board (10 by 10 inches)
- ☐ Clear packaging tape
- ☐ Dixie cup
- ☐ Pencil
- ☐ Indelible black marker

WHAT TO DO

Using the clear packaging tape, neatly tape together the three squares of foam-core board to form three sides of a cube.

Using the Dixie cup as a template, draw five circles on the inner side of the bottom board, four circles on the inner side of the left-hand board, and one circle on the inner side of the right-hand board—just like the pattern of dots on a die. Color each dot solid black with the indelible black marker.

Place the inverted die on a table or the floor, stand across the room, and stare at it.

WHAT HAPPENS

This is an optical illusion: although the object is concave, your brain, accustomed to seeing dice as cubes, perceives the object as convex. Moving a few feet to your left or right makes the die appear to rotate in the direction opposite to your motion. If you have an assistant hold the die in the palms of his hands and then move the die in a circular motion, the die seems to float above his hands and turn in the opposite direction of that motion.

WHY IT WORKS

Changes in visual perception occur with a constant stimulus pattern when two alternate figures can be perceived against the same ground. The reversible figure also induces spontaneous oscillation in visual perception. As you stare at the shape, the two shapes seem to alternate back and forth more rapidly. Shifting which spot on the die you stare at will help determine which of the two alternatives you will see.

BIZARRE FACTS

- The word *bones* is slang for dice, which were originally carved from bones.
- Ancient Greek philosopher Plato said God invented dice.
- In the New Testament, when Jesus is crucified, Roman soldiers cast dice for his clothes.

- In 1598, Queen Elizabeth issued a search and seizure order against the manufacturers of loaded dice.
- In his play *The Merchant of Venice*, William Shakespeare wrote:

  "If Hercules and Lichas play at dice
  Which is the better man, the greater throw
  May turn by fortune from the weaker hand."

- Physicist Albert Einstein is frequently quoted as having said "God does not play dice with the universe."
- The slang expression "no dice" denotes a negative response or failure.
- On their 1972 album *Exile on Main Street*, the Rolling Stones recorded the song "Tumbling Dice."

Dicey Vision

If you stare at the shape in this experiment, you will perceive one of the two possible visual patterns. Eventually, your mind reaches physiological satiation and inhibits you from perceiving this first visual organization, allowing your mind to perceive the second form. Your mind gradually becomes satiated with the second form, and the first form reappears, and this oscillation continues with increased frequency.

■ Thirty-six possible combinations of numbers can be rolled with a pair of dice. The odds of rolling a 7 are one out of six. The odds of rolling a 6 or 8 are five out of thirty-six. The odds of rolling a 5 or 9 are one out of nine. The chances of rolling a 4 or 10 are one out of twelve. The odds of rolling 3 or 11 are one out of eighteen. The odds of rolling a 2 or 12 are one out of thirty-six.

Electric Wiggle Tube

WHAT YOU NEED

- ☐ X-acto knife
- ☐ Large cardboard box
- ☐ 3-foot-long flexible plastic tube (¼ inch in diameter)
- ☐ Black paint
- ☐ Paintbrush
- ☐ Water
- ☐ Saran Wrap (3 inches square)
- ☐ Rubber band
- ☐ Electrical tape
- ☐ Desk lamp

WHAT TO DO

With adult supervision, use the X-acto knife to cut a small hole for the plastic tube to fit through in one end of the narrow side of the box, near the base.

Paint the inside and outside of the cardboard box black. Let dry.

Fill a tall glass with water, immerse one end of the tube in the water, and suck the other end to fill the tube with water. Put the square of Saran Wrap

over one end of the tube and secure in place with a rubber band to make a watertight seal.

Holding the tube with the open end higher than the sealed end (to prevent water from leaking out of the tube), insert the tube through the hole in the cardboard box so that the open end of the tube sticks outside the box by one inch. Use electrical tape to seal the space between the tube and the hole.

Bend the tube around inside the box and position the sealed end upward. Shine the desk lamp into the open end of the tube, and turn off the other lights in the room. If necessary, cover the box and peek inside.

WHAT HAPPENS

The tube of water lights up like a tubular bulb.

WHY IT WORKS

The light travels through the water, is reflected by the inside wall of the tube, and continues to bounce along the length of the tube, illuminating the water-filled tube—in the same way light travels through optical fibers.

BIZARRE FACTS

- Light entering the water core is reflected off the inside wall of the tube if the light hits the wall at a critical angle (around 82 degrees to the perpendicular or right angle). Light hitting the inside wall of the tube at less than the critical angle leaks out.
- An optical fiber is an outer glass sleeve filled with a glass core.
- Fiber-optic cables contain numerous thin, clear glass fibers, each carrying a pulsed (digital) light signal. Communications equipment (telephones, computers, and televisions) transforms electronic signals into digital light signals, and then turns them back into electronic signals when they are received at the other end of the line. Each optical fiber can carry several thousand telephone calls at once, transmitting enormous amounts of information.
- In his poem "The Charge of the Light Brigade," English poet Lord Tennyson glorified the courage of the British troops during the Crimean War (1853-1856) with the frequently

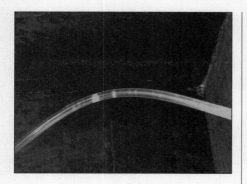

quoted line: "Theirs not to reason why,/Theirs but to do and die." The charge by the 673 men of the Light Brigade was a suicidal bloodbath. When the Russians retreated from the port city of Balaklava, taking captured Turkish artillery with them, the British commander sent one of his men, a Captain Nolan, to order the British brigade of light calvary to stop the Russians from taking the Turkish cannons. Nolan miscommunicated the order, directing the light bri-gade to prevent the Russians from leaving with *any* cannons—Turkish or Russian. The resulting battle left one hundred thirteen British soldiers dead and another 134 wounded or captured.

■ The name Lucifer means "light-bearer" in Latin. However, Lucifer is not the name for the devil. The name Lucifer appears only once in the Bible: "How art thou fallen from heaven, O Lucifer, son of the morn-ing" (Isaiah 14:12). The verse refers to an arrogant king of Babylon who intended to ascend to heaven to set his throne above the stars of God. Saint Jerome (circa 342–420 C.E.)—noting that Isaiah uses the epithet "day star," that Satan is seen to "fall like lightning from Heaven" (Luke 10:18), and that the Latin word *Lucifer* means "light-bearer"—incorrectly interpreted the verse to mean that Satan and Lucifer were one and the same. English poet John Milton popularized this misinterpretation in 1667 in his epic poem *Paradise Lost,* in which Satan is also named Lucifer.

Enlightening Edison

Thomas Edison did not invent the lightbulb. In 1802, English chemist Sir Humphrey Davy invented an arc light, by making a platinum wire glow by passing an electric current through it. Edison patented his carbon-filament lamp in 1879. Edison became known as the father of the lightbulb by building local power plants to generate and distribute electricity.

Electricity Generator

WHAT YOU NEED

- ☐ 10 feet of uninsulated 8-gauge copper wire
- ☐ Drinking glass
- ☐ 12-inch plank of pinewood (¾ by 6 inches)
- ☐ Ruler
- ☐ Pencil
- ☐ Two-hole metal strap
- ☐ Two ¾-inch screws
- ☐ Two 8-inch lengths of pinewood (¾ by 1½ inches)
- ☐ Two 4-inch lengths of pinewood (¾ by 1½ inches)
- ☐ Electric drill with ¹⁄₁₆-inch bit and Phillips head screwdriver bit
- ☐ Four 1-inch drywall screws
- ☐ Two eyelet screws
- ☐ 6-inch wooden dowel (⁵⁄₁₆-inch diameter)
- ☐ Ten round magnets with hole in center
- ☐ Two thumbtacks
- ☐ String or dental floss
- ☐ Galvanometer

WHAT TO DO

Wrap the wire tightly around the drinking glass, making approximately sixty coils 3 inches in diameter and leaving 1 foot of wire at both the beginning and the end. Slide the coil from the glass, and sew the free ends of the wire around the coil to hold the wires together firmly and compactly.

Measure the 12-inch plank of pinewood to find the center. Position the coil vertically in the center so it splits the board into two 3-inch halves. Fasten the coil to the plank with the two-hole metal strap and two ¾-inch screws. Use foam or cardboard under the loop if necessary to help hold the coil firmly in place.

Hold one 4-inch length of pine to the end of an 8-inch length to form an L. Drill a hole through the end of the piece. Screw the pieces together with a drywall screw. Repeat to form a second L.

Drill a hole in the center of the free end of the 4-inch piece of an L. Screw an eyelet screw into the hole. Repeat with the second L.

Turn each L upside down, center it along the narrow edge of the 6-inch plank, drill a hole through the end and into the base, and screw the pieces together with a drywall screw. Repeat with the second L.

Insert the dowel through the circular magnets, and push a thumbtack into each end of the dowel.

Cut two pieces of string each 8 inches long, and with each, tie one end to a dowel rod thumbtack and the other end to the eyelet screw at the end of the L above it so that the magnets and dowel create a swing through the center of the wire coil.

Connect the two free ends of the wire coil to the galvanometer.

Slowly swing the magnets and dowel back and forth horizontally through the coil of copper wire.

WHAT HAPPENS

As the magnet swings through the coil of wire, the galvanometer registers electrical current. When the magnet stops swinging, the galvanometer records zero electrical current. The faster you swing the magnet through the coils, the steadier the flow of electricity the galvanometer registers.

WHY IT WORKS

The movement of magnetic lines of force across a wire creates a current in the wire. Rapid movement of the magnet produces a stable and usable flow of electricity called alternating current (created by the alternating movement of the magnet).

BIZARRE FACTS

- In 1820, Danish scientist Hans Oersted discovered that passing an electric current through a wire produces magnetism—creating the electromagnet.
- In 1831, English chemist and physicist Michael Faraday discovered that the opposite was also true—moving a magnet through a coil of copper wire produces an electric current that flows through the wire. His published findings introduced the principle of "electromagnetic induction." American physicist Joseph Henry actually discovered electromagnetic induction before Faraday but failed to publish his findings.
- In a copper wire attached to both poles of a battery, the free electrons in the wire flow from the negative to the positive pole, producing an electric current that travels in the opposite direction of the electron flow.
- Faraday's discovery of "electromagnetic induction" paved the way for building massive generators to create electricity to power everything from electric lights to machinery, launching the industrial revolution.
- The farad, a unit of electrical capacitance, is named after Michael Faraday.

Riding the Wave

In 1873, Scottish scientist James Clerk Maxwell narrowed the laws of electricity and magnetism down to four basic mathematical formulas—based on the premise that electromagnetic fields travel in waves near the speed of light. Maxwell also theorized that electric circuits could produce electromagnetic waves that travel at the speed of light. He also contended that light was just one of many forms of electromagnetic waves, which was proven in 1888 when German physicist Heinrich Hertz discovered radio waves.

Flaming Nondairy Creamer

WHAT YOU NEED

- ☐ Clean, empty coffee can with plastic lid
- ☐ Drill with a ¼-inch bit
- ☐ Plastic drinking straw
- ☐ Scissors
- ☐ Electrical tape
- ☐ Candle (1-inch diameter)
- ☐ Matches
- ☐ 2 tablespoons Coffeemate powdered nondairy creamer

WHAT TO DO

With adult supervision, drill a hole in the side of the coffee can, as close to the bottom rim as possible.

Insert 1 inch of the straw into the hole in the side of the can. Use the scissors to cut small pieces of electrical tape to seal the spaces between the straw and the hole in the can.

Using the scissors, cut off a piece of candle 1 inch long, making sure to leave enough wick so you can light it later. Using the matches, light the bigger candle and carefully let ten drops of hot wax drip to the center of the inside of the metal bottom of the coffee can. Secure the shorter candle upright inside the coffee can on the center of the bottom.

Outdoors, place 2 tablespoons of nondairy creamer around the candle inside the coffee can. With adult su-

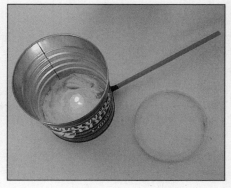

pervision, carefully light the candle and cover the coffee can with the lid.

Blow hard into the straw. Remember to be careful with fire!

WHAT HAPPENS
The plastic lid blows off the coffee can with a momentary blast of fire and smoke.

WHY IT WORKS
Powdered nondairy creamer, when suspended in air, is flammable.

BIZARRE FACTS
■ Flour is also highly flammable when suspended in air, explaining why explosions often occur in grain storage facilities.

■ Genghis Khan's armies carried rations of dried milk.
■ Early Africans milked buffaloes.
■ Powdered milk is made by evaporating all the water from pasteurized skimmed milk.

Things Aren't Always What They Seem to Be

Some brand-name nondairy creamers contain casein, which is a dairy product.

Floating Bubbles

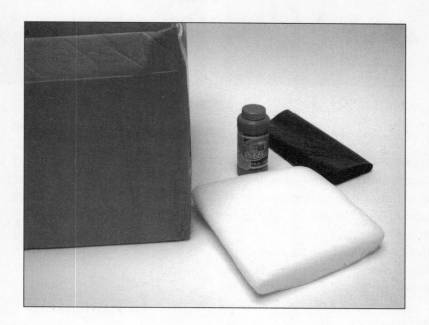

WHAT YOU NEED

☐ Cardboard box
☐ Plastic garbage bag
☐ Work gloves
☐ **Large block of dry ice (always handle dry ice with work gloves)**
☐ Bottle of bubbles and blow ring

WHAT TO DO

Line the inside of the box with the garbage bag, and, wearing work gloves, place the block of dry ice in the bottom of the box. (Never touch dry ice with your bare hands since it can burn your skin if it touches it directly. Also never eat dry ice as it can be fatal.) Blow some bubbles into the box.

WHAT HAPPENS

The bubbles float in midair, get larger, absorb each other, and change colors.

WHY IT WORKS

The bubbles float on a cushion of carbon dioxide. The carbon dioxide is a heavy gas, and the bubbles are filled with air, which is made up of much lighter gases. The bubbles change colors and turn transparent as the soap film falls to the bottom of the

bubble—eventually causing the bubble to pop.

BIZARRE FACTS

- Dry ice is frozen carbon dioxide gas.
- When dry ice melts, the frozen carbon dioxide returns to gas form rather than liquid form.
- Dry ice is colder than ice made from frozen water. Dry ice can reach a temperature as low as minus 112 degrees Fahrenheit.
- To make dry ice, carbon dioxide gas is compressed into a liquid and then cooled and evaporated to make carbon dioxide snow, which is then compressed into blocks of solid dry ice.
- If ingested, dry ice can cause death because of its low temperature.

- Dry ice can be used to ship perishable foods because the ice does not melt.
- Inserting a small piece of dry ice into an uninflated balloon will produce enough carbon dioxide to inflate the balloon.

Boiling Ice

A small piece of dry ice, placed in a glass of water, will appear to make boiling water which steams. This produces an excellent "boiling cauldron" effect for Halloween celebrations.

Flying Ping-Pong Ball

WHAT HAPPENS

The Ping-Pong ball hovers upward in the middle of the airstream and stays in place, as if floating on a cushion of air. You can also tilt the blow-dryer roughly 20 degrees and the Ping-Pong ball will remaining floating.

WHY IT WORKS

In 1738, Swiss mathematician Daniel Bernoulli discovered "Bernoulli's principle"—namely, that the pressure in a fluid drops as it moves faster. The Ping-Pong ball stays floating in the column of fast-moving air because the high pressure surrounding the column pushes the ball back into the column of low-pressure air.

BIZARRE FACTS

■ In the early 1900s, manufacturers frequently promoted multiple uses for a newly devised electrical appliance in the hopes of increasing sales of the product. An early advertisement for a vacuum cleaner called the Pneumatic Cleaner showed a woman using a hose connected to the machine to dry her hair, assuring readers that the vacuum produced a "current of pure, fresh air from the exhaust." At the time, using the exhaust from a vacuum

WHAT YOU NEED

☐ Blow-dryer
☐ Ping-Pong ball

WHAT YOU DO

With adult supervision, plug in the blow-dryer, turn it on "high cool," and aim the nozzle straight up in the air. Hold a Ping-Pong ball directly over the nozzle and gently release the ball.

cleaner was the only way to blow-dry hair. The handheld electric hair dryer had yet to be invented because no one had developed a motor small enough to power it—until 1922, when the advent of the blender produced the first fractional horsepower motor.

■ In 1920, two companies in Racine, Wisconsin—the Racine Universal Motor Company and Hamilton Beach—essentially combined the vacuum cleaner with the blender to create the handheld hair dryer. The Racine Universal Motor Company introduced the "Race"; Hamilton Beach launched the "Cyclone."

■ While the first handheld hair dryers were bulky and frequently overheated, manufacturers made improvements during the decade that followed, providing adjustable speeds and temperature settings. In 1951, Sears and Roebuck introduced a portable handheld dryer that attached to a pink plastic cap that fit over the woman's hair.

■ In 1968, Leandro P. Rizzuto, founder of Continental Hair Products in New York City, developed the hot comb, and three years later, introduced the first handheld pistol-grip blow-dryer to the United States.

More Hot Air

The development of the blow-dryer enabled hair salons to accomplish in minutes what used to take hours; thus they handled more clients in less time, increasing revenues.

Freaky Fire Extinguisher

WHAT YOU NEED

☐ Drinking glass

☐ Candle

☐ Saucer

☐ Matches

WHAT TO DO

With adult supervision, use a match to melt a few drops of wax from the candle to adhere the candle to the saucer. Place the drinking glass next to the saucer. Light the candle.

Standing on the side of the drinking glass away from the candle, blow air toward the side of the glass.

WHAT HAPPENS

The candle flame is extinguished as if you blew air directly through the glass.

WHY IT WORKS

In 1738, Swiss mathematician Daniel Bernoulli first observed that whenever air moves, its pressure drops. When you blow air at the glass, the rapidly moving stream of air lowers the air pressure along its path. The stronger air pressure around it keeps the stream of lower air pressure close to the drinking glass, rather than being diffracted away by the glass.

BIZARRE FACTS

■ Another way to demonstrate Bernoulli's principle is to cut two long narrow strips from a newspaper. Suspend the two strips an inch apart from one another and blow a stream of air between the ends of the strips. The ends will come together rather than moving apart as you might expect.

■ Bernoulli's principle explains how airplanes fly. Air streams over the top of the curved wings faster than air flows beneath the wings, creating lower pressure above the wing and causing the plane to become airborne. The plane engines thrust the plane forward fast enough to create enough low air pressure to cause lift.

■ Ancient Greek philosopher Empedocles incorrectly claimed that everything in the universe is composed of four elements (earth, air, fire, and water), which he insisted were bonded together by love and driven apart by strife.

■ In an episode of the television comedy show *Mad About You,* avant-garde artist Yoko Ono urges Paul Reiser to make a documentary film about the wind, prompting Paul to go crazy trying to film wind.

Ghost in a Glass

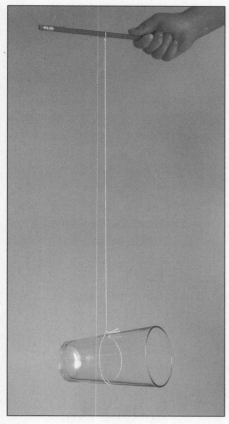

ing glass and the other end of the string tightly around the center of a pencil.

Hold one end of the pencil, allowing the drinking glass to be suspended in air by the string.

Roll the pencil very slowly between your fingers.

WHAT HAPPENS

The drinking glass emits a faint knocking or tapping sound.

WHY IT WORKS

When you begin turning the pencil, the string begins to wrap around the pencil—until the weight of the glass pulls the string back to its starting position. This small movement causes the string to vibrate slightly, and the glass amplifies this mysterious tap.

BIZARRE FACTS

■ You can use this Ghost in a Glass experiment to conduct an eerie séance. Tell your friends that the glass is a conduit to the spirit world. Have your friends ask the glass a question, and by secretly rolling the pencil, you can manipulate the glass to rap once for yes and twice for no.

WHAT YOU NEED

☐ Kite string (18 inches long)
☐ Clean, empty drinking glass
☐ Pencil

WHAT TO DO

Tie one end of the piece of string tightly around the center of the drink-

- Sound travels at different speeds depending on the temperature and the medium it is passing through. The denser and more compressible the medium is, the slower the sound travels. Under standard atmospheric conditions, sound travels at 742 miles per hour at sea level when the temperature is 32 degrees Fahrenheit. Sound travels faster at higher altitudes and at higher temperatures.

- Sound travels about four times faster through water than air and about fifteen times faster through steel. Sound travels approximately 1,130 feet per second through air, roughly five thousand feet per second through water, and nearly twenty thousand feet per second through steel. Sound travels faster through solids because the molecules are closer together.

- In the first century C.E., Roman science writer Gaius Pliny determined that light travels faster than sound based on his observation that he could see lightning strike before he heard the sound of the accompanying thunder.

- Since sound travels through air at approximately 1,130 feet per second, you can determine how far away a bolt of lightning has struck by counting the number of seconds between the flash of lightning and the sound of thunder. For every five seconds you count, the storm is one mile away.

- Ancient Greek philosopher Aristotle theorized that sound cannot be heard in a vacuum. In 1654, German inventor Otto von Guericke proved Aristotle's theory by creating a vacuum in a jar containing a bell. When rung, the bell made no noise. Sound cannot travel in a vacuum because there are no molecules to vibrate the sound.

- High-pitched sounds and low-pitched sounds travel at the exact same speed.

- Sounds pitched higher than can be heard by the human ear are called "ultrasound." Sounds pitched lower than can be heard by the human ear are called "infrasound."

- The human ear cannot hear sounds pitched above 20,000 Hz (waves per second), but bats and dolphins can hear ultrasounds with pitches up to 120,000 Hz.

- High-energy ultrasound "guns" can shatter steel girders.

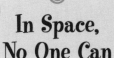

In Space, No One Can Hear You Scream

In most science-fiction movies, the rockets on spacecraft make violent roaring sounds and explosions as they travel through space. In reality, spacecraft do not make any sounds in space, since space is a vacuum.

Heavy Wood

WHAT YOU NEED

☐ Kitchen or postal scale

☐ A block of wood (2 by 4 by 6 inches)

☐ A plastic container

☐ Water

WHAT TO DO

Weigh the block of wood on the scale and record the weight.

Place the empty plastic container on the scale and record the weight.

Fill the plastic container three-quarters full of water. Record the combined weight of the plastic container and water. Add the weight of the block of wood to the weight of the plastic container of water.

Place the block of wood in the water-filled container and record the combined weight. Note that the combined weight registered on the scale equals the previously calculated weight of the wood, container, and water.

With your fingertips, press the block of wood until it is just submerged under the surface of the water in the container. Record the weight.

WHAT HAPPENS

The total weight is greater than the previously combined total of the wood, container, and water.

WHY IT WORKS

Ancient Greek mathematician Archimedes discovered that the volume of an object immersed in a tub of water is equal to the volume of the water it displaces. The block of wood in this experiment floats on the water. The scale records the force required to push the buoyant wood underwater to displace enough water to equal its volume.

BIZARRE FACTS

■ Archimedes' principle states that an object placed in a liquid seems to lose an amount of weight equal to the weight of the fluid it displaces.

■ Archimedes also concluded that a floating object displaces an amount of liquid equal to its own weight.

■ Archimedes observed that an object placed in a fluid is buoyed upward with a force equal to the weight of the fluid it displaces.

■ Archimedes invented the catapult, the Archimedean screw (a device for raising water, consisting of a screw fitted within a cylindrical case), and a system of mirrors designed to concentrate the sun's rays to set enemy ships on fire.

Fuzzy Math

When the Romans captured Syracuse on the island of Sicily in 212 B.C.E., the Roman commander, Marcellus, ordered his soldiers not to harm the Greek mathematician Archimedes, who lived in the city. The Roman soldiers, however, failed to recognize Archimedes and killed him.

Homemade Electroscope

WHAT YOU NEED

- ☐ Clean, empty glass jar
- ☐ Cardboard
- ☐ Pencil
- ☐ Scissors
- ☐ Wire clothes hanger
- ☐ Pliers
- ☐ Elmer's Glue-All
- ☐ Aluminum foil
- ☐ Electrical tape
- ☐ Plastic ballpoint pen
- ☐ Piece of silk fabric

WHAT TO DO

Place the glass jar upside down on the cardboard and trace around it with the pencil to create a circle. With the scissors, cut out the circle.

With a pair of pliers, snip a 6-inch length of wire from the clothes hanger. Bend ¾ inch of one end of the wire to form an L-shape.

Push the straight end of the wire through the center of the cardboard

disk so the straight end protrudes 2 inches. Fill the hole with glue. Let dry.

Cut a piece of foil into a rectangular strip ½ inch wide and 3 inches long. Fold the rectangular strip in half and, using the glue, adhere it over the bent end of the wire.

Cut a second piece of foil into a 12-inch square, and roll the square of foil into a tight ball. Push the aluminum ball onto the straight end of the wire.

Place the bent end of the wire inside the glass jar until the cardboard disk caps the jar like a lid. Using electrical tape, tape the cardboard disk securely to the rim of the jar.

Rub the body of the plastic ballpoint pen with the piece of silk fabric for three minutes, then wave the pen closely around the foil ball.

WHAT HAPPENS

The two strips of foil inside the jar fly apart from each other.

WHY IT WORKS

Rubbing the plastic pen with the piece of silk fabric gives the body of the pen a positive charge of static electricity. Waving the positively charged pen over the ball causes negative charges in the wire to move upward toward the foil ball, leaving the strips of foil with a positive charge. The two positively charged strips then repel each other.

BIZARRE FACTS

■ Static electricity is an electrical charge that cannot move through an object, but instead remains static. The object holds the static electrical charge—until it is touched by another object that conducts electricity.

■ Around 2000 B.C.E., the Chinese discovered how to make silk from silkworm cocoons. For nearly three thousand years, the Chinese kept this discovery a secret. Poor Europeans, unable to afford genuine silk, would make imitation silk by beating cotton with sticks to soften the

A Shocking Discovery

In one of the most dangerous experiments in history, American statesman Benjamin Franklin proved that lightning is static electricity by attaching a key to the string of a kite and then flying the kite in a thunderstorm. The wet string conducted the electricity to the key, which emitted sparks.

fibers and then rubbing the fabric against a large stone. They called the resulting shiny cotton "chintz." The word *chintzy* came to mean cheap or inferior quality.

- Ancient Greeks discovered that rubbing a piece of amber (fossilized tree resin) with a cloth causes the amber (now charged with static electricity) to attract feathers. The word *electricity* stems from the Greek word for amber.

- Rubbing two objects together can produce static electricity. For instance, rubbing a plastic comb on a wool sweater charges the comb with static electricity—transferring electrons from the wool to the comb and giving the comb a negative charge.

- When you comb your hair quickly on a dry day, your hair loses electrons and becomes positively charged, while the comb gains electrons and becomes negatively charged. As you comb your hair, the static electricity generated makes it crackle.

- When you walk across a carpet, you generate static electricity. If you touch a metal object, like a doorknob, the positive charge you have generated will leap to the uncharged doorknob—creating a spark and giving you a slight shock.

Homemade Water Blaster

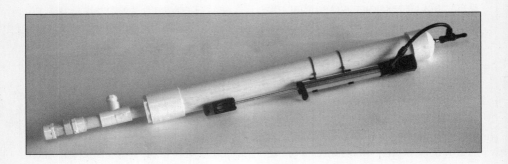

WHAT YOU NEED

- ☐ Electric drill with ⅜-inch bit and ¹⁄₁₆-inch bit
- ☐ 2-inch-diameter PVC end cap
- ☐ Metal automotive tire valve with nuts to tighten (available at automotive supply store)
- ☐ Adjustable wrench
- ☐ Utility knife
- ☐ 2-foot-long 2-inch-diameter PVC pipe
- ☐ 2-inch-diameter PVC coupler
- ☐ PVC pipe reducer fitting from 2-inch diameter to ¾-inch diameter
- ☐ Two 4-inch lengths of ¾-inch-diameter PVC pipe
- ☐ Three-way PVC coupler with two ¾-inch openings and one ½-inch female screw opening
- ☐ ½-inch screw plug
- ☐ PVC pipe adapter (from ¾ inches to garden hose)
- ☐ Teflon tape
- ☐ Garden hose shutoff valve
- ☐ ¾-inch PVC end cap
- ☐ PVC pipe primer
- ☐ PVC pipe glue
- ☐ Bicycle pump with attachment nozzle
- ☐ Four plastic multipurpose ties (12 inches long)

WHAT TO DO

With adult supervision, use the drill and ⅜-inch drill bit to drill a hole in the center of the end cap. Unscrew the tightening nut from the metal automotive tire valve, insert the tire valve through the inside of the end cap (making sure the rubber gasket covers the drilled hole), and tighten the nut in place with the adjustable wrench.

Using the utility knife, carefully scrape off any burrs left from the ends of the cut pieces of PVC pipe to assure a clean seal when the pipes are securely sealed together airtight.

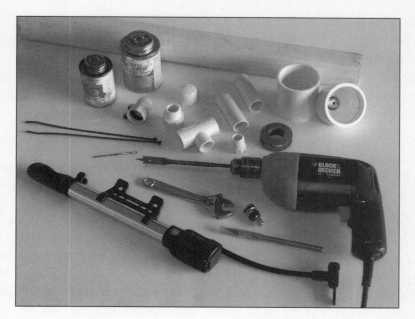

Coat the ends of each PVC pipe and the inside of each PVC coupler, the outside and inside of the PVC pipe reducer fitting, and end cap with PVC pipe primer. Using the PVC pipe glue, attach the 2-inch-diameter PVC end cap (previously prepared with the tire valve) to one end of the 2-foot-long 2-inch-diameter PVC pipe. To the other end of the 2-foot-long 2-inch-diameter PVC pipe, glue the 2-inch-diameter PVC coupler. Glue the PVC pipe reducer fitting inside the 2-inch diameter PVC coupler.

Glue one of the 4-inch lengths of ¾-inch-diameter PVC pipe into the ¾-inch hole of the PVC pipe reducer fitting. Glue the free end of the ¾-inch-diameter PVC pipe into one of the ¾-inch openings of the three-way PVC coupler. Screw the ½-inch screw plug loosely into the ½-inch opening in the three-way PVC coupler. Glue the remaining 4-inch length of ¾-inch-diameter PVC pipe into the remaining ¾-inch hole of the three-way PVC coupler.

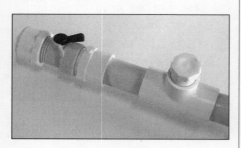

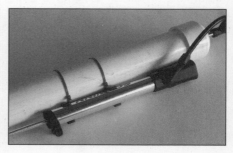

Glue the smooth end of the PVC pipe adapter (from ¾ inches to garden hose) to the free end of the 4-inch length of ¾-inch-diameter PVC pipe. Using Teflon tape around the grooves, attach the garden hose shutoff valve to the adapter.

Using the electric drill with the ⅟16-inch bit, drill a hole through the center of the ¾-inch PVC end cap. Using Teflon tape, screw the end cap to the end of the garden hose adapter.

Attach the nozzle of the bicycle pump to the tire valve, and using the multipurpose plastic ties, secure the body of the bicycle pump to the water chamber.

Turn the garden hose shutoff valve closed, unscrew the ½-inch screw plug, and submerge the entire device underwater until the chamber fills with water. Screw the plug back in place.

Pump air into the chamber with ten strokes of the bicycle pump.

Aim the nozzle at your target (keep the nozzle lower than the chamber), and flip the switch on the garden hose shutoff valve to on.

Never aim at a person's face.

WHAT HAPPENS

The water blasts through the nozzle, shooting about twenty feet.

WHY IT WORKS

When you open the valve, the air pressure stored in the chamber escapes, forcing the water out through the small nozzle; since the nozzle is so small, the water escapes with great force.

BIZARRE FACTS

- In 1988, Lonnie Johnson, an aerospace engineer from Los Angeles, California, invented the first squirt gun to utilize air pressure in its design. He named his invention the "Power Drencher," but three years later, after receiving a patent, Johnson renamed the toy "Super Soaker" and launched a nationwide advertising campaign.
- Lonnie Johnson, inventor of the Super Soaker, holds a master's degree in nuclear engineering from Tuskegee University, served as a captain in the Air Force (where he received astronaut training as a Space Shuttle mission specialist, and worked at the Jet Propulsion Laboratory in Pasadena, California (where he worked on the Galileo and Mars Observer projects and the Cassini mission to Jupiter).
- The pressurized water from a Super Soaker can be used to unclog a bathroom drain.
- Water is the most common compound on earth.
- Less than 3 percent of the water on earth is freshwater. The remaining 97 percent is salt water.

- The average human being drinks eight thousand gallons of water during his or her lifetime.
- The designers of the "unsinkable" RMS *Titanic* divided the ship into sixteen watertight compartments, which could be individually sealed shut by the flip of an electronic switch. If any two of the compartments flooded, the ship would remain afloat. In the unlikely event that three compartments flooded, the ship would sink. When the *Titanic* hit an iceberg on its maiden voyage just before midnight on April 14, 1912, the resulting 300-foot-long gash opened six of the sixteen compartments. Also, the ship's designers provided only sixteen lifeboats—the minimum number for a 15,000-ton vessel (the *Titanic* weighed 46,000 tons and carried more than 2,200 passengers and crew). At 2:20 A.M. on April 15, 1912, the *Titanic* sank. Only 705 people survived—less than one-third of those aboard.
- In 1978, Random House published *Woman's Day Crockery Cuisine,* a cookbook that included a recipe for "Silky Caramel Slices." The recipe instructed readers to heat an unopened can of evaporated milk in a Crock-Pot, but accidentally neglected to tell readers to fill the pot with water. When Random House discovered that following the recipe could cause the can of evaporated milk to explode, it recalled ten thousand copies of the cookbook.
- The 1995 movie *Waterworld,* made for a record $175 million, bombed at the box office. In the film, the earth is completely submerged underwater, yet all the people, living in small boats on the water in the sun, are covered with dirt and grease.

Lake Lake, I Presume

In 1859, Scottish explorer David Livingstone reached what is currently known as Lake Malawi and asked the local people what it was called. When they told him *nyasa,* he named the huge body of water Lake Nyasa. Unbeknownst to Livingstone, the word *nyasa* meant "mass of waters," so the explorer actually named the body of water "Lake Lake." In 1964, the newly independent Malawi government renamed the lake Lake Malawi.

Ice Cream Machine

WHAT YOU NEED

- ☐ Mixing bowl
- ☐ Box of instant pudding (to make six ½-cup servings)
- ☐ 3 cups milk
- ☐ Whisk
- ☐ Small, clean, empty coffee can with lid (net weight 13 ounces)
- ☐ Electrical tape
- ☐ Large, clean, empty coffee can with lid (net weight 39 ounces)
- ☐ Ice
- ☐ Rock salt

WHAT TO DO

Empty the contents of the pudding mix into the mixing bowl. Add 3 cups milk (according to the directions on the back of the box), and mix well using the whisk.

Pour the pudding solution into the small, clean coffee can. Secure the plastic lid in place and use the electrical tape to make the lid watertight.

Place the small coffee can into the large coffee can. Fill the rest of the

large can with ice up to the top of the small can. Fill the rest of the space with rock salt. Secure the plastic lid in place and use the electrical tape to make the lid watertight.

Take the can outside and roll it across the lawn or patio for fifteen minutes. Bring the can back inside, peel the tape from the lid of the large can, pour out the melted ice and salt, and refill with fresh ice and fresh salt. Secure the lid in place again, and roll the can outside for another fifteen minutes.

Bring the can back inside, peel off the tape from the larger can, pour out the melted ice and salt, and wash off the smaller can with tap water in the sink. Dry the can. Store in the freezer for twelve hours. Peel off the tape from the smaller can, remove the lid, and scoop the contents into bowls.

WHAT HAPPENS
You've made ice cream.

WHY IT WORKS
Mixing ice with salt in the compartment around the small can creates freezing temperatures, causing the mixture inside the small can to freeze. Rolling the large can causes the smaller can to roll around in the ice. As the smaller can rolls, air bubbles are whipped into the ice cream, increasing the volume of the mix.

BIZARRE FACTS
■ No one knows who first invented ice cream or when. In the late 1500s, Europeans used ice, snow, and saltpeter to freeze mixtures of cream, fruit, and spices.

■ Almost all ice cream was made at home until 1851, when Baltimore milk dealer Jacob Fussell established the first ice cream factory.

- The edible ice cream cone, invented by Italo Marchiony of Hoboken, New Jersey, was first served at the 1904 World's Fair in St. Louis, Missouri.
- The most popular flavor of ice cream in the United States is vanilla, accounting for approximately one-third of all the ice cream sold in the country. The second most popular flavor is chocolate, followed by strawberry.
- The United States produces more than one billion gallons of ice cream, ice milk, sherbet, and water ice every year.
- Approximately 10 percent of all the milk produced in the United States is used to make ice cream and other frozen desserts.
- As mayor of Carmel, California, actor Clint Eastwood's first act was to legalize ice cream parlors.

- Americans eat more ice cream than do the people of any other nation in the world.
- The average American eats roughly 14.5 quarts of ice cream in a year.

You Scream, I Scream, We All Scream for Ice Cream

On July 24, 1988, Palm Dairies Ltd. of Alberta, Canada, created the world's largest ice cream sundae—made from 44,689 pounds, 8 ounces of ice cream; 9,688 pounds, 2 ounces of syrup; and 537 pounds, 3 ounces of toppings.

Inflating Glove

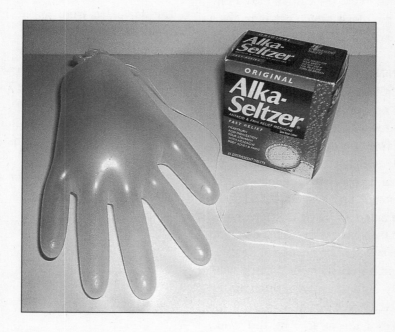

WHAT YOU NEED

☐ Dental floss
☐ Latex glove
☐ Water
☐ Four Alka-Seltzer tablets

WHAT TO DO

Tie a small loop in the end of a 12-inch-long piece of dental floss. Thread the free end of the dental floss through the loop, creating a slipknot large enough to encircle the open end of the glove. Set aside.

Carefully fill the four fingers of the glove with water, leaving the thumb empty. Carefully slip four Alka-Seltzer tablets inside the dry thumb. Place the prepared slipknot over the open end of the glove, then pull the end of the dental floss to tighten the knot and securely close the end of the glove. Wrap the dental floss tightly around the slipknot several times, then tie another knot to secure in place.

Hold the thumb of the glove and guide each one of the four Alka-Seltzer tablets into its own water-filled finger.

WHAT HAPPENS

The glove inflates like a balloon.

WHY IT WORKS

When activated in water, the Alka-Seltzer tablet releases carbon dioxide gas, filling the sealed glove. With nowhere to go, the carbon dioxide inflates the glove, which expands like a balloon.

STRANGE FACTS

- Right-handed people live an average of nine years longer than left-handed people do.
- Paul McCartney and Ringo Starr, the only two left-handed Beatles, are the only two surviving Beatles.
- German Kaiser Wilhelm II frequently attempted to conceal his withered arm by posing with his hand resting on a sword or by holding gloves.
- When terrorists sent letters filled with anthrax spores through the United States Postal Service in the wake of the attacks on the World Trade Center and Pentagon on September 11, 2001, sales of latex gloves skyrocketed.
- During the Middle Ages, throwing down a glove signaled a challenge to a duel and was called "throwing down the gauntlet." Whoever picked up the glove accepted the challenge.
- Soldiers from every country in the world salute with their right hand.
- The Japanese word *karate* means "empty hand."
- The three major candidates in the 1992 presidential election—George Bush, Bill Clinton, and Ross Perot—were all left-handed.
- The word *glove* originates from the Anglo-Saxon word *glof*, meaning "palm of the hand."
- William Shakespeare's father was a glovemaker.

All You Need Is Glove

During the Middle Ages, a knight would attach a woman's glove to his helmet to demonstrate love or devotion.

Instant Silver

WHAT YOU NEED

- ☐ A metal cake pan
- ☐ Aluminum foil
- ☐ Water
- ☐ Measuring spoons
- ☐ Baking soda
- ☐ Cooking thermometer
- ☐ Tarnished silver

WHAT TO DO

Line the bottom of the metal cake pan with a sheet of aluminum foil and fill with enough water to cover the silverware (roughly 2 inches). Add 2 tablespoons baking soda per quart of water. Using the cooking thermometer, heat the water above 150 degrees Fahrenheit. Place the tarnished silverware in the pan so it rests on the aluminum foil. Do not let the water boil. Let the silverware soak over the heat for ten minutes, then turn off the heat and let the water cool before removing the silverware.

WHAT HAPPENS

The silver comes out sparkling clean—without any scrubbing.

WHY IT WORKS

Silver gradually darkens because silver chemically reacts with sulfur-containing substances in the air to form silver sulfide, which is black. In this experiment, a chemical reaction converts the silver sulfide back into silver—without removing any of the silver. Like silver, aluminum forms compounds with sulfur—but with a greater affinity than silver. In this experiment, sulfur atoms are transferred from silver to aluminum, freeing the silver metal and forming aluminum sulfide. The warm baking soda solution carries the sulfur from the silver to the aluminum, creating aluminum sulfide, which adheres to the aluminum foil or forms tiny yellow flakes in the bottom of the pan.

BIZARRE FACTS

- The silver and aluminum must be in contact with each other during this experiment because a small electric current flows between them during the reaction.
- Polishing silverware with an abrasive cleanser removes the silver sulfide and some of the silver from the surface. Other chemical tarnish removers dissolve the silver sulfide, but also remove some of the silver.
- New Jersey is home to a spoon museum featuring over 5,400 spoons from every state and nearly every country.
- Sodium bicarbonate, more commonly known as baking soda, was originally used as an ingredient in cake batter to make cakes rise; hence the combination of the words *baking* and *soda*.

- Baking soda has an almost unlimited shelf life.
- Baking soda was used to clean the Statue of Liberty for the 1976 Bicentennial celebration.
- A box of baking soda can be found in nine out of ten refrigerators. According to the *Los Angeles Times*, "More refrigerators are likely to have baking soda than working lightbulbs." Baking soda chemically neutralizes odors by turning into a sodium salt and giving off water and carbon dioxide.
- When mixed in cake batter and heated, baking soda releases a carbon dioxide gas that causes the cake to rise.
- Baking soda cleanses by neutralizing fatty acids found in most dirt and grease.
- Baking soda has been used to reduce air pollution in factory smokestacks. The baking soda, when pulverized to an appropriate particle size, is, like other sodium sorbents, one of the most effective collectors of sulfur dioxide. Injecting sorbent-grade sodium bicarbonate directly into the flue gas ducts of coal-fired boiler systems desulfurizes flue gas. The baking soda reacts with sulfur dioxide to form sodium sulfate, and the cleaned flue gas exits through the stack.
- Baking soda has also been used to increase the effectiveness of sewage treatment plants. Baking soda helps maintain proper pH and

alkalinity in biological digesters, fostering trouble-free operation of both anaerobic and aerobic treatment plants. Used in maintenance doses, baking soda boosts sludge compaction, alkalinity, and methane gas production while reducing biological oxygen demand and controlling sulfide odors. Plus, it's environmentally safe.

■ The United States Environmental Protection Agency and the Navy's Civil Engineering Lab have jointly developed an inexpensive way to use baking soda to decontaminate polluted soil.

■ Baking soda can be used to increase the butterfat content of cow and goat milk. Adding baking soda to cow and goat feed increases the pH in the animals' rumina, lowering the acidity, making for a more favorable environment for the microbacteria that aid digestion. This causes the animals to eat more, increasing milk production and the butterfat content of the milk.

Splish Splash

Baking soda can restore lakes damaged by acid rain. In 1985, Cornell professor James Bisongi, Jr., restored Wolf Pond, a virtually dead fifty-acre lake in the Adirondacks, by adding nearly twenty tons of baking soda to the water to dramatically reduce the acidity.

Jell-O Garden

WHAT YOU NEED

☐ Paper towel
☐ Scissors
☐ Saran Wrap
☐ Sunflower seeds (uncooked)
☐ Ziploc bag
☐ 1 box Jell-O
☐ Water
☐ Clean plastic drinking cups

WHAT TO DO

Cut a 3-inch-wide strip from a sheet of paper towel, dampen, and lay on top of a strip of Saran Wrap. Place the seeds on top of the paper towel at the intervals recommended on the seed packet. Cover with another 3-inch-wide strip of damp paper towel, then roll the paper and plastic together, place in a Ziploc bag, and store in a warm place (on a windowsill, on top of a refrigerator).

When the seeds begin to sprout, mix up Jell-O with water according to the instructions and pour the gelatin into several clear plastic drinking cups

O. Once the flower begins to grow, plant the Jell-O and the seedling in a pot of soil or outside in your garden.

BIZARRE FACTS

■ Cranberry Jell-O is the only flavor of Jell-O derived from genuine fruit, rather than artificial flavoring.

■ To sow small seeds evenly and easily, mix one package of Jell-O with just enough water to create a thick gel with the consistency of mustard. Mix small seeds into the gel, pour the mixture into a clean, empty Ziploc bag, seal securely, cut a small hole in the corner of the bag, and then squeeze the bag to squirt a line of goop from the hole into even rows. You can also squirt a line of the goop onto one-inch-wide strips of paper towel, let dry, then plant the paper strips in the garden.

■ To give tomato plants additional nitrogen, mix an envelope of powdered Jell-O into one cup of boiling water, stir until the gelatin powder dissolves, mix with three cups of cold water, then apply around the base of the plant.

■ Work a few teaspoons of powdered Jell-O into the soil of houseplants to absorb water and prevent it from leaking out of the bottom of the pot. The absorbent gelatin also reduces how often you need to water the plants.

■ Mary Wait, the Jell-O inventor's wife, came up with a name for the fruit-flavored animal jelly by com-

and let harden in the refrigerator. Place the cups of hardened gelatin on the counter, remove the Ziploc bag, unroll the Saran Wrap, and plant several sprouting sunflower seeds into the center of the surface of the Jell-O cups, pushing them down approximately the length of your finger. Let the cups of gelatin stand on the counter near the window to get light. Do not water. Observe.

WHAT HAPPENS

The seeds grow in the Jell-O, enabling you to observe the root structures and the seedling sprouts. The Jell-O may also grow a thin, harmless mold on the surface.

WHY IT WORKS

The gelatin keeps the seeds moist and helps the plants retain water, the nitrogen in Jell-O enhances plant growth and hastens sprouting, and the sugar feeds microbes, producing more nutrients for the plant. The seedling draws moisture from the Jell-

bining the word *jelly* with *-O,* a popular suffix added to the end of a slew of food products at the time.

- Gelatin, a colorless protein derived from the collagen contained in animal skin, tendons, and bone, is extracted by treating hides and bone with lime or acid. The material is then boiled, filtered, concentrated, dried, and ground into granules, which dissolve in hot water and congeal into a gel when the solution cools.

- As a food supplement, gelatin supplies the body with several amino acids lacking in wheat, barley, and oats.

- According to the *1993 Guinness Book of Records,* the world record for the largest Jell-O is held by Paul Squire and Geoff Ross, who made a 7,700-gallon watermelon-flavored pink Jell-O in a tank supplied by Pool Fab on February 5, 1981, at Roma Street Forum in Brisbane, Australia.

- A box of Jell-O can be found in three out of four American pantries.

- Americans eat over 690,000 boxes of Jell-O on an average day.

Jell-O Spies

In July 1950, the FBI arrested thirty-two-year-old electrical engineer Julius Rosenberg as a spy for the Soviet Union. According to the FBI, Rosenberg had torn a Jell-O boxtop in half, given a piece to his brother-in-law, David Greenglass, and told him that his contact at Los Alamos would produce the other half. The contact turned out to be spy courier Harry Gold, who received atomic-energy data from Greenglass and paid him five hundred dollars, allegedly giving the Soviet Union the secret of the atom bomb. Although Rosenberg insisted on his innocence, he and his wife, Ethel, were sentenced to death in 1951, and after several appeals, in June 1953, the Rosenbergs became the first Americans ever executed for using Jell-O.

Kaleidoscope

WHAT YOU NEED

- [ ] Ruler
- [ ] Pencil
- [ ] Cardboard mailing tube 2 inches in diameter with plastic end caps (available at an office supply store)
- [ ] Serrated knife
- [ ] Three pieces of mirror, each 1½ by 10 inches (available at a glass company)
- [ ] Electrical tape
- [ ] Paper towels
- [ ] Acetate sheet (from the cover of a clear folder)
- [ ] Indelible marker
- [ ] Scissors
- [ ] Rubber cement
- [ ] 14-ounce frosted plastic drinking cup (available at the supermarket)
- [ ] Colored plastic drinking straws
- [ ] Electric drill with ½-inch bit
- [ ] Sandpaper
- [ ] Wrapping paper
- [ ] Clear laminating plastic

WHAT TO DO

Measure a 10½-inch length of the mailing tube, draw a circle around the tube as a guideline, and with adult su-

pervision, use a serrated knife to cut the length from the mailing tube.

Using electrical tape, carefully cover all four edges of each of the three pieces of mirrored glass to prevent cuts from the sharp edges. Avoid taping the mirrored surfaces.

Carefully place the mirrors together to form a triangle, with the mirrored surfaces on the inside of the triangle. Making certain the edges are even, tape the mirrors together along the edges and then run strips of tape around the entire triangle.

Slide the triangle of mirrors into the mailing tube, aligning one end of the triangle with one end of the tube—leaving ½ inch of free space at the other end of the tube.

Crumple up small pieces of paper towel and, using the pencil, shove them between the walls of the mailing tube and outsides of the mirrors to cushion the mirrors and wedge them snugly inside the tube.

Lay a plastic end cap from the mailing tube on the sheet of acetate, and use the indelible marker to trace a circle around the end cap. Using a pair of scissors, cut one disk from the acetate.

On the end of the mailing tube that is even with the mirrors, paint rubber cement along the cardboard edge of the mailing tube. Let dry for a few minutes, then apply a second coat of rubber cement. Fit the acetate disk over the end of the tube, press tightly, then stand the tube on end so the disk is pressed against the floor or tabletop, and let dry.

Using scissors, cut the bottom from the 14-ounce frosted plastic cup roughly ¼ inch above the base (so that the piece can fit over one end of the mailing tube like a cap and be secured in place). Cut ¼-inch pieces from the

colored drinking straws. Place the straw pieces into the plastic cup bottom. Fit the piece over the acetate disk on the end of the mailing tube. Hold the tube up to the light, and look through the other end while turning the tube. If you do not see changing designs, add more pieces of colored straws.

When you are content with the number of straws in the cup, remove the cup, apply rubber cement to make a ¼-inch band around the perimeter of the mailing tube, and replace the plastic cup, holding firmly until the glue sets.

With adult supervision, use the electric drill to make a ½-inch hole in the center of the plastic end cap of the mailing tube. Use fine sandpaper to smooth the edges of the hole.

Cut a circle 1 inch in diameter from the sheet of acetate, and glue the disk over the hole on the inside of the plastic end cap. Place the prepared end cap over the open end of the mailing tube to create a peephole.

Cut a piece of wrapping paper to fit around the tube, and adhere to the tube with rubber cement. Then cover the tube with clear laminating plastic.

Look through the peephole, hold the other end of the kaleidoscope up to the light, and gently turn the tube.

WHAT HAPPENS
You see changing symmetrical patterns of bright color.

WHY IT WORKS
The colors at the end of the tube are reflected in the mirrors, creating multiple images. Turning the tube shifts the colored shapes, creating new patterns.

BIZARRE FACTS
▓ Sir David Brewster invented the kaleidoscope and patented his invention in 1817.
▓ Designers have used kaleidoscopes to find new patterns for fabrics, rugs, and wallpapers.
▓ Most mirrors consist of a thin sheet of silver or aluminum attached to a sheet of high-quality glass.

- In the children's story "Snow White and the Seven Dwarves," as told by the Brothers Grimm, the wicked queen asks her magic mirror, "Mirror, mirror, on the wall, who's the fairest of them all?"
- In Lewis Carroll's novel *Through the Looking-Glass*, the sequel to *Alice's Adventures in Wonderland*, Alice steps through a mirror above a fireplace into a parallel universe inhabited by strange creatures including Tweedledee, Tweedledum, and Humpty Dumpty.
- In the song "Lucy in the Sky with Diamonds," the Beatles sing about a girl named Lucy with kaleidoscope eyes.

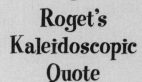

Roget's Kaleidoscopic Quote

In 1818, Dr. Peter M. Roget (who published his thesaurus in 1852) praised the kaleidoscope in *Blackwood's Magazine*: "In the memory of man, no invention, and no work, whether addressed to the imagination or to the understanding, ever produced such an effect."

Marble Run

WHAT YOU NEED

- ☐ Ruler
- ☐ Pencil
- ☐ 10-inch length of pinewood (1 by 1 inch)
- ☐ Electric drill with ¼-inch bit
- ☐ Sandpaper
- ☐ Four identical drinking glasses
- ☐ Karo light corn syrup
- ☐ Corn oil
- ☐ Red wine vinegar
- ☐ Shampoo
- ☐ Four identical marbles

WHAT TO DO

Using the ruler and pencil, mark off every 2 inches along the 10-inch length of wood. With adult supervision, drill a ¼-inch hole through the center of each mark in the wood, so you have created four holes. Use sandpaper to smoothen the wood.

Line up the four drinking glasses in a row. Fill each glass with a different liquid to the same level, roughly 1 inch from the lip of the glass. Rest the 10-inch length of wood across the tops of the glasses so that one long edge of the wood sits over the center of each

glass, with a hole positioned near the center of each glass.

Rest a marble over each of the four holes on the wood. Tilt the wood toward the center of the drinking glasses so the four marbles roll into the glasses at the same time and from the same height. Observe the marbles.

WHAT HAPPENS

The four marbles sink to the bottom of the liquids at different speeds. The liquid containing the marble that hits the bottom first is the least viscous.

WHY IT WORKS

Different liquids possess different viscosities. The longer a marble takes to sink in a liquid, the higher that liquid's viscosity.

BIZARRE FACTS

- For an interesting variation on this experiment, gently warm all four liquids equally to see how temperature affects viscosity.
- As you pour corn syrup, it forms coils due to its high viscosity.
- Viscosity is a measure of a liquid's

thickness or resistance to flow. The thicker a liquid, the higher the viscosity. Viscosity is a type of internal friction.

- Pushing an object into a viscous liquid increases the viscosity, making the liquid resist more.

Mutual Attraction

In liquids, the molecules, while attracted to each other, are attracted less strongly than the molecules in solids and can move around. Freezing temperatures turn liquids into solids because the lower temperature causes the molecules to move slower and attract each other. Hot temperatures turn liquids into gases because the molecules move faster and have less attraction to each other.

Matchbook Rocket

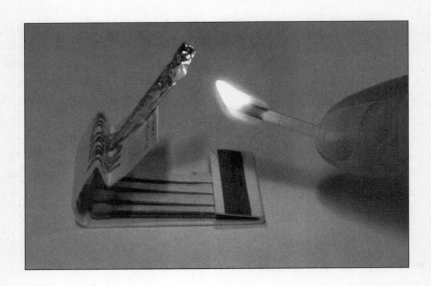

WHAT YOU NEED

- ☐ Aluminum foil
- ☐ Scissors
- ☐ Ruler
- ☐ Book of paper matches
- ☐ Toothpick

WHAT TO DO

With adult supervision, use the scissors to cut a rectangle 1 by 2 inches from the aluminum foil. Fold the rectangle in half to form a 1-inch square.

Remove one paper match from the matchbook and hold a toothpick evenly alongside it. Place the folded rectangle of foil over the head of the top two-thirds of the match and toothpick, wrap the foil securely around both sticks, and slide out the toothpick, leaving a narrow open channel leading to the match head.

Open the matchbook and fold the cover in half by bringing the bottom edge of the cover to the top of the matchbook. Then fold that free top edge back up to form a small ledge on the matchbook cover.

Stand the prepared match on the launch pad, foil end up. Carefully light a second match and hold the flame beneath the foil-wrapped head of the first match.

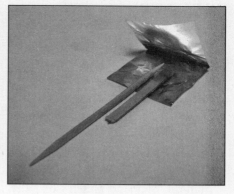

WHAT HAPPENS

When the match head ignites under the foil, the match goes flying across the room.

WHY IT WORKS

Gas escaping through the channel propels the match, in accord with Sir Isaac Newton's third law of motion: for every action there is an opposite and equal reaction.

BIZARRE FACTS

- In 1680, British physicist Robert Boyle coated a piece of paper with the element phosphorus and coated the tip of a small stick of wood with sulfur, creating the first chemical match. Phosphorus, however, was a rare and costly element, making the wide-scale production of matches economically infeasible.

- As early as the seventeenth century, British children played a game in which a lit candle was placed in the middle of a room and the players tried to jump over it without extinguishing the flame. The game inspired the nursery rhyme, "Jack be nimble, Jack be quick, Jack jumped over the candlestick."

- In 1826, British pharmacist John Walker, attempting to invent a new explosive, invented the first friction match by accident. When a glob of one of his chemical solutions dried on the end of his mixing stick, he tried to remove it by rubbing the coated end of the stick against the stone floor. The tip immediately burst into flame. The solution was a mixture of antimony sulfide, potassium chlorate, gum, and starch. Walker never patented his invention.

- The first commercial matches, manufactured by Samuel Jones in London, England, in the late 1820s, were called Lucifers and boosted the popularity of smoking tobacco.

- In 1830, French chemist Charles Sauria developed phosphorus matches, inadvertently giving rise to "phossy jaw," a fatal disease technically known as phosphorus necrosis and primarily affecting factory workers. Highly poisonous, phosphorus essentially causes bones to disintegrate and deform. The phosphorus scraped from the heads of a pack of matches could also be used to commit suicide or murder.

- Mice chewing on phosphorus match heads at night frequently caused kitchen fires.
- In 1896, a brewery ordered more than fifty thousand matchbooks from the Diamond Match Company, originating the idea of advertisements on the covers of matchbooks and creating the demand for machinery to mass-produce matches.
- In 1911, the Diamond Match Company introduced the first nontoxic match, waiving the right to a patent so other companies could produce the matches as well, eliminating the scourge of "phossy jaw."

- During World War II, the United States military airdropped millions of matchbooks printed with propaganda messages behind enemy lines to boost the morale of people living in countries occupied by the Axis powers.

Close before Striking

In 1892, Joshua Pusey, a lawyer from Lima, Pennsylvania, invented the matchbook—foolishly placing the striking surface inside the front cover, enabling the slightest friction to cause the entire matchbook to go up in flames.

Mayonnaise Madness

room temperature and let it defrost. Shake the jar for five minutes.

WHAT HAPPENS

In the freezer, the mayonnaise turns into large chunks of white curds floating in yellow oil. When left to defrost, the mayonnaise remains chunks of white curds. No matter how hard you shake the jar, the mayonnaise remains flakes of white curds in yellow oil.

WHY IT WORKS

When frozen, mayonnaise separates and curdles because the emulsion breaks down. As mayonnaise freezes, the oil globules expand and break away from the film that surrounds them. The oil congeals and rises to the surface, and no amount of mixing can make the mayonnaise look appetizing again.

WHAT YOU NEED

☐ Jar of mayonnaise
☐ Freezer

WHAT TO DO

Peel off the paper label from the mayonnaise jar. Place the jar of mayonnaise in the freezer overnight. In the morning, remove the jar from the freezer. Place the jar on a counter at

BIZARRE FACTS

■ Mayonnaise is a combination of the word *Mahón,* a town on the island of Minorca, and *-aise,* the French suffix for *-ese.*

■ In the eighteenth century, the Duc de Richelieu discovered a Spanish condiment made of raw egg yolk and olive oil in the port town of

Mahón on the island of Minorca, one of the Balearic Islands. He brought the recipe for "Sauce of Mahón" back to France, where French chefs used it as a condiment for meats, renaming it *mayhonnaise*. When mayonnaise arrived in the United States in the early 1800s, it was considered a haute French sauce, too difficult to prepare. The invention of the electric blender and the advent of bottled dressings catapulted mayonnaise into the mainstream as a sandwich spread.

- In 1912, Richard Hellmann, a German immigrant who owned a delicatessen in Manhattan, began selling his premixed mayonnaise in one-pound wooden "boats," graduating to glass jars the following year. Hellmann's eventually extended its distribution from the East Coast to the Rocky Mountains. Meanwhile, Best Foods, Inc., had introduced mayonnaise in California, calling it Best Foods Real Mayonnaise and expanding distribution throughout the West. Eventually the two companies merged under the Best Foods, Inc., banner, but since both brands of mayonnaise had developed strong followings, neither name was changed.
- Hellmann's Real Mayonnaise and Best Foods Real Mayonnaise are essentially the same, although some people find Hellmann's mayonnaise slightly more tangy.
- Mayonnaise is not salad dressing. Real mayonnaise contains at least 65 percent oil by weight, vinegar, and egg or egg yolks. Salad dressings contain a minimum of 30 percent oil, at least 4 percent egg yolk, cooked starch, water, and vinegar or other specified acid ingredients.
- Mayonnaise provides essential fatty acids that the body cannot manufacture and vitamin E. It also aids in the absorption of vitamins A, D, E, and K.
- An unopened jar of commercial mayonnaise is microbiologically safe to eat for an indefinite period of time—as is an opened jar that is refrigerated.

The Penicillin of Champions?

The acid ingredients in mayonnaise—vinegar and lemon juice, along with salt—destroy many bacteria. Studies at the University of Wisconsin have shown that commercial mayonnaise retards the growth of salmonella in salads inoculated with the bacterium.

Milk Paint

WHAT YOU NEED

- [ ] 1½ cups Carnation Nonfat Dry Milk powder
- [ ] ½ cup water
- [ ] Food coloring or pigment (available at an art supply store)
- [ ] Blender
- [ ] Paintbrush

WHAT TO DO

Mix 1½ cups Carnation Nonfat Dry Milk and ½ cup water until it is the consistency of paint. Blend in your choice of food coloring or pigment until you attain the desired hue. Thin the paint by adding more water, thicken the paint by adding more powdered milk. Brush on as you would any other paint. Let the first coat dry for at least twenty-four hours before adding a second coat. Let the second coat dry for three days.

WHAT HAPPENS

The milk is paint.

WHY IT WORKS

The milk is an emulsion that suspends the pigment.

BIZARRE FACTS

- Early American colonists made their milk paint from the milk used to boil berries, resulting in an attractive gray color.
- Milk paint is extremely durable.
- To strip milk paint, apply ammonia, and allow it to dry for about four days, then apply bleach. Make sure you are stripping the paint in a well-ventilated area.
- A cow gives nearly 200,000 glasses of milk in her lifetime.
- As early as the Stone Age, people mixed pigments in animal fat to make waterproof paints.
- Pigments are ground from minerals, earths, plants, and animals. For instance, ultramarine is ground from the semiprecious stone lapis lazuli, vermilion is ground from mercuric sulfide from volcanic rocks, ocher is ground from iron oxide, and brazilwood is made from shavings of Brazilwood.

- The ancient Egyptians developed paints made from egg whites mixed with pigment. Artists during the Italian Renaissance used paints made from ground pigments suspended in linseed oil.
- In tomb paintings, the ancient Etruscans depicted women in white and men in red.
- During his lifetime, Vincent Van Gogh sold only one painting: *Red Vineyard at Arles*.
- While traveling to the South Sea Islands, French impressionist Paul Gauguin worked as a laborer on the Panama Canal.

Old MacDonald Had Some Paint

Barns are traditionally painted red because early farmers mixed milk paint with blood from slaughtered animals to achieve the red color.

Model Camera

WHAT YOU NEED

- ☐ A tall, empty tissue box
- ☐ Ruler
- ☐ Pencil
- ☐ Scissors
- ☐ Cardboard tube (from a roll of toilet paper)
- ☐ Electrical tape
- ☐ Wax paper
- ☐ Scotch tape
- ☐ Magnifying glass

WHAT TO DO

Using the ruler and pencil, mark off a ⅜-inch wide lip along the perimeter of the top panel of the tissue box. Using the scissors, cut out this square panel from the top of the tissue box.

On the bottom of the tissue box, position the cardboard tube in the center of the panel and draw around it to make a circle. With the scissors, cut out the circle. Insert the cardboard tube into the resulting hole, allowing 2 inches of the cardboard tube to stick out from the hole. Use electrical tape to secure the tube in position and prevent any light from getting into the box from the cracks between the tube and the hole.

Cut out a square of wax paper that is ½ inch larger in both height and width than the square cut from the tissue box. Place this sheet of wax paper over the hole in the top of the tissue box and tape in place with Scotch tape.

Aim the cardboard tube at an object and look at the wax paper screen.

Hold the magnifying glass in front of the cardboard tube and move it back and forth to bring the object into focus on the wax paper screen.

WHAT HAPPENS

The model camera forms an image upside down on the tracing paper, the

Blow It Out Your Nose

In 1914, the Kimberly-Clark Company of Neenah, Wisconsin, developed Cellucotton, a cotton substitute used by the United States Army as surgical cotton during World War I. After the war, Kimberly-Clark, eager to find a commercial market for Cellucotton, introduced "Kleenex Kerchiefs" as a disposable facial towel to remove makeup. Housewives soon discovered that their husbands were using their Kleenex Kerchiefs as disposable handkerchiefs, and that's the way Kimberly-Clark has marketed Kleenex tissues ever since.

same way a genuine camera does (but without making a photograph).

WHY IT WORKS

The magnifying glass acts like a lens. Light waves travel in a straight line from the subject to the magnifying glass. When the light waves go through the cardboard tube, they project an upside-down image on the wax paper. The image is inverted because light travels in a straight line, causing the light from the top of the subject to strike the bottom part of the wax paper. Light from the bottom of the subject strikes the top part of the wax paper.

BIZARRE FACTS

■ The human eye works much like a camera. Light enters the lens of the human eye, which forms the image on the retina upside down. Cells in the retina send signals along the op-

tic nerve to the brain, which interprets the image rightside up.

■ In 1936, Kimberly-Clark inserted into every box of Kleenex tissues a list of forty-eight alternative uses for the product, including polishing furniture, cleaning pots and pans, and cleaning car windshields.

■ The trademarked brand name Kleenex passed into the American vernacular as a synonym for "a disposable paper tissue" and can be found as a word in the dictionary.

Musical Wineglasses

WHAT YOU NEED

- ☐ 3-foot-long piece of pinewood ($\frac{3}{4}$ by 6 inches)
- ☐ Green felt (1 by 4 feet)
- ☐ Staple gun and staples
- ☐ Scissors
- ☐ Two strips of $\frac{1}{4}$-inch wood (1 inch by 3 feet)
- ☐ Two strips of $\frac{1}{2}$-inch-thick foam (3 feet by 1 inch) (available at a fabric or craft store)
- ☐ Elmer's Carpenter's Wood Glue
- ☐ Electric drill with $\frac{1}{16}$-inch bit
- ☐ Eight identical wineglasses
- ☐ Screwdriver
- ☐ Six $\frac{3}{4}$-inch drywall screws
- ☐ Measuring cup
- ☐ Water
- ☐ Food coloring
- ☐ Vinegar
- ☐ Small bowl

WHAT TO DO

Wrap the felt around the wooden board and use the staple gun to secure the felt to the back of the board, wrapping the corners neatly. Using the scissors, trim any excess felt so the board rests level on a tabletop.

Glue a foam strip along the length of one side of each one of the wooden strips. Let dry.

With adult supervision, drill a $\frac{1}{16}$-inch hole in the center and 1 inch from both ends of both foam-backed wooden strips.

Line the glasses in a row on the felt-covered board, spacing each glass 1 inch apart. Place the foam-backed wooden strips lengthwise on the felt-covered board, one strip on each side of the row of glasses—just slightly overlapping the base of each glass, with the foam-backed side facing down.

Use drywall screws to attach the wooden strips to the felt-covered board to hold the glasses in place.

Fill the first wineglass with 1 ounce of water, the second glass with 2 ounces of water, the third glass with 3 ounces of water, and continue this progression until you fill the last glass with 8 ounces of water. (You may need to add more or less water to each glass to create a tuned one-octave scale.) Use the food coloring to color each glass of water a different color.

Pour 2 ounces of vinegar into the bowl, dip the index finger of each hand into the vinegar, and then rub your finger around the rims of the wineglasses in a circle to create sounds.

WHAT HAPPENS
By rubbing your fingers along the rims of the glasses, you create a strange, ringing tone in each glass.

WHY IT WORKS
The friction from your finger causes the glass to vibrate and the resulting longi-tudinal sound waves resonate verti-cally in the glass itself and also travel around the circumference of the glass. The more water in the glass, the lower the frequency (or note) of the sound waves traveling through the glass. The ability to create different pitches en-ables us to tune different glasses to dif-ferent musical notes on the scale, creating a musical instrument.

BIZARRE FACTS
■ For an interesting variation on this experiment, fill long-necked bottles with increasing amounts of water, then blow over the bottles to make musical sounds, creating a pan flute.
■ Different pitches of an instrument are actually different frequencies of sound. High-pitched sounds are high-frequency wavelengths, mean-ing the sound waves travel closer together and are more numerous. Low-pitched sounds are low-frequency wavelengths, meaning the sound waves travel further apart and are less numerous.
■ The same note sounds different on different instruments because each instrument produces its own unique set of vibrations.
■ In 1887, American inventor Thomas Alva Edison made the first sound recording. He recorded his voice reciting the words "Mary had a little lamb."
■ The eardrum is a thin membrane of skin approximately ⅖ inch in diam-eter stretched taut across the audi-

tory canal like the skin of a drum. The eardrum is so thin that even the tiniest sound wave vibrates it to and fro. As the eardrum vibrates, it swings a tiny bone called the hammer across another bone called the anvil. The anvil shakes a third bone called the stirrup, amplifying the vibrations to the cochlea, coiled tubes resembling a snail shell, filled with liquid, and lined with minute hairs. Sound waves traveling through the fluid bend the hairs which in turn stimulate tiny nerve fibers that send the signals along the cochlear nerve to the brain, which interprets the impulses as sounds.

■ A cricket's eardrums can be seen on the side of each front leg.

■ The fleshy, curved part of the ear on the outside of the head is called the auricle.

■ The auricles on an African elephant measure up to four feet wide, making them the largest auricles of any mammal.

■ The stirrup bone in your ear is the smallest bone in the body, tinier than a grain of rice.

■ William Faulkner got the title for his 1929 novel *The Sound and the Fury* from a line in William Shakespeare's *MacBeth* (act 5, scene 5):

"Life is but a walking shadow, a poor player
That struts and frets his hour upon the stage,
And then is heard no more; it is a tale
Told by an idiot, full of sound and fury,
Signifying nothing."

To Russia, with Love

To save money in the 1970s, the United States State Department hired Soviet construction workers to build a new United States embassy in Moscow. The State Department, expecting the Soviets to plant listening devices in the building, planned to remove them upon taking control of the building. When the eight-story multimillion-dollar building was completed in 1985, United States security experts detected thousands of electronic diodes mixed into the concrete and hundreds of tiny microphones with their wires hidden inside steel beams and reinforcing rods. Rather than dismantling the building to find all the bugs, Congress budgeted 240 million dollars to build a new one—this time, without Soviet construction workers.

Mystery Bottle

WHAT YOU NEED

☐ Clean, empty 1-liter plastic soda bottle
with cap

☐ Water

☐ Blue food coloring

☐ Clean glass mixing bowl

WHAT TO DO

Fill the bottle with water, add five drops blue food coloring, screw on the cap securely, and shake well to mix up the coloring. Remove the cap.

Place the bottle upside down inside the bowl. A little water will come out from the bottle. Raise the bottle 1 inch.

Observe. Raise the bottle another inch. Observe again. Raise the bottle yet another inch. Observe.

WHAT HAPPENS

The level of the water in the bowl will rise only to the height of the mouth of the bottle. The rest of the water remains inside the bottle and does not pour out into the bowl.

WHY IT WORKS

The force pushing down on the surface of the water in the bowl (atmospheric pressure) is greater than the

force pushing the water downward in the neck of the bottle, preventing the water from escaping from the bottle. When the mouth of the bottle is raised above the surface of the water in the bowl, air gets into the bottle, allowing water to escape. This is the theory behind pet feeders that allow water to run from a reservoir as the pet drinks water from a bowl.

BIZARRE FACTS

- Temperature affects air pressure. Hot air expands, creating low air pressure. Cold air contracts, causing high air pressure.
- The weight of air from the top of the atmosphere to the layers below creates air pressure. At sea level, the air pressure is 14.7 pounds per square inch. At the top of Mount McKinley (an altitude of 20,320 feet), the air pressure is 6.8 pounds per square inch.

- Drinking through a straw utilizes air pressure. Sucking on the straw creates reduced air pressure inside the straw, and the greater air pressure pushing down on the surface of the liquid in the drinking glass or bottle pushes the liquid up through the straw.
- Ancient peoples made the first bottles from animal skins. Archaeologists believe that people discovered how to blow bottles from molten glass on the end of a hollow iron pipe in the first century B.C.E.

The Friendly Skies

In 1978, to prevent employees from stealing the miniature liquor bottles served aboard flights, Pan Am secretly installed a gadget in the liquor cabinets aboard its planes to count the number of times the cabinet was opened. Stewardess Susan Becker, discovering one of the unfamiliar gadgets aboard an airborne Boeing 707 and thinking it might be a bomb, informed the pilot who made an emergency landing, instructing the eighty passengers to use the emergency exits with the inflatable slides—at a cost of approximately twenty thousand dollars.

In April 1986, reporter Geraldo Rivera hosted a two-hour live national broadcast from the basement of the Lexington Hotel in Chicago, where a sealed-off cement-walled room, believed to be the secret vault of gangster Al Capone, would be opened. The vault, blasted open after an hour and a half, contained two empty bottles.

In 1989, wine merchant William Sokolin, having paid $300,000 for a 1787 bottle of Margaux once owned by Thomas Jefferson, held up the bottle before three hundred wine collectors at the Four Seasons restaurant in New York City, where he hoped to sell it for $519,750. Sokolin accidentally dropped the bottle and it smashed to pieces.

Parachute

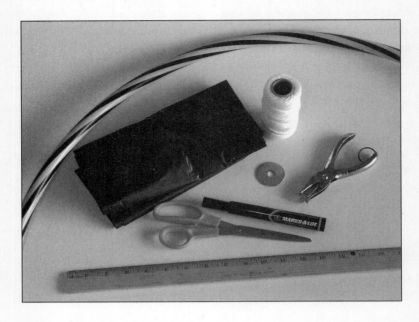

WHAT YOU NEED

- ☐ Large lawn trash bag (available at a hardware store)
- ☐ Scissors
- ☐ Hula Hoop
- ☐ Indelible marker
- ☐ Hole puncher
- ☐ Four 2-foot lengths of string
- ☐ Metal washer (1 inch in diameter)

WHAT TO DO

Using the scissors, cut the trash bag open along one of the side seams and along the bottom seam. Place the open trash bag on a flat surface, and lay the Hula Hoop on top of the sheet of plastic. With the indelible marker, trace around the Hula Hoop to make a circle. Use the scissors to cut out the circle of plastic.

Using the hole puncher, punch a hole in the plastic 1 inch in from the edge of the circle. Thinking of the first hole as twelve o'clock on a clockface, punch a second hole at three o'clock, a third hole at six o'clock, and a fourth hole at nine o'clock.

Tie one end of a 2-foot length of string through each hole. Tie the free

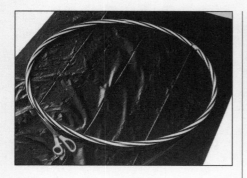

ends of the four strings together through the washer.

Fold up the plastic parachute, wrap the string around it, and toss it up into the air.

WHAT HAPPENS
As the parachute descends, the plastic spreads and the washer floats gently back to earth.

WHY IT WORKS
As you toss the folded parachute upward, the compact mass goes far without encountering much air resistance. When gravity pulls the washer and parachute back to earth, the parachute opens and the plastic sheet encounters air resistance, slowing the descent.

BIZARRE FACTS
- In 1495, Leonardo da Vinci designed a triangular parachute that looks like a tent.
- In 1783, French physicist Sebastian Lenormand made the first successful parachute jump from a tower.

- In 1797, French inventor André Jacques Garnerin made the first parachute jump from a balloon. He made the jump over the Parc Monceau in Paris.
- Parachutes used for humans measure from twenty-four to twenty-eight feet across when extended.
- Parachutes fall approximately fifteen feet per second, depending on the weight of the parachutist.
- A person wearing an open parachute hits the ground at roughly the same speed as if jumping from a height of ten feet.

Look before You Leap

A successful parachute jump requires a height of at least five hundred feet above the ground, otherwise the parachute does not have enough distance to open.

- On April 24, 1967, as the manned space capsule of the Soviet spacecraft *Soyuz 1* returned to earth, the straps of the parachute snapped. The capsule plummeted to the ground, killing lone Russian cosmonaut Vladimir Komarov.
- In April 1981, a United States Air Force pilot, instructed to shoot down an unmanned target plane in a peacetime training mission over Florida, got confused and shot down a manned F-4 Phantom fighter jet instead. The two crewmen parachuted to safety, but the $3.3 million F-4 Phantom fighter crashed in the Gulf of Mexico.

Ping-Pong Ball and Marble Mystery

WHAT YOU NEED

☐ Clean, empty mayonnaise jar with lid

☐ Water

☐ Marble

☐ Ping-Pong ball

WHAT TO DO

Fill the mayonnaise jar with water. Place the marble and the Ping-Pong ball in the jar and seal the lid securely. Lay the jar on its side and spin. Observe what happens to the marble and the Ping-Pong ball when the jar stops spinning.

WHAT HAPPENS

The marble rolls to one end of the jar while the Ping-Pong ball moves to the center of the jar.

WHY IT WORKS

An object that sinks in fluid in a gravitational field will also be affected by a centrifugal force applied to that fluid. A spinning bottle creates a centrifugal force field that causes the marble to move from the center of the jar to one of the two ends of the jar. The Ping-Pong ball, however, unaffected by the gravitational field of the fluid, floats to the top of the fluid and seeks the region of zero centrifugal field (the center, at zero radius).

BIZARRE FACTS

- Marbles can be played for fun, returning the marbles to each original owner at the end of the game, or as a gambling game called "keepsies," in which each player keeps the marbles won.
- The phrase "losing your marbles," derived from playing "keepsies," means losing your mind.
- Almost every ancient culture independently developed the game of marbles.
- Primitive man used chestnuts, hazelnuts, and olives as marbles.
- Ancient Egyptian children played marbles with semiprecious stones as early as 3000 B.C.E.
- As early as 1435 B.C.E., children on the Greek island of Crete played with marble balls made from jasper and agate.
- The word *marble* is derived from the Greek word *marmaros* (meaning polished white agate).
- Ancient Romans made clear glass marbles from silica and ash.
- Ancient Saxons and Celts used ordinary stones and pellets of clay as marbles.
- The 1560 painting *Children's Games* by Flemish painter Pieter Brueghel depicts children shooting marbles.
- Although marbles have been made from clay, stone, wood, glass, and steel, most marbles today are made from glass.
- The record number of hits in one minute during a Ping-Pong game is 172, held by Thomas Busin and Stefan Renold of Switzerland, during a game held on November 4, 1989.
- Table tennis developed in England during the late 1800s.
- Regulation Ping-Pong balls must measure from 1.4 to 1.5 inches in diameter and weigh from $\frac{1}{12}$ ounce to $\frac{1}{11}$ ounce.
- The first popular arcade video game duplicated a Ping-Pong game and was called "Pong."

Balloon Mystery

If a passenger sitting in the backseat of a car holds a string attached to a helium balloon, when the car accelerates forward, the balloon floats forward (toward the front seat), not backward. When the car rounds a curve, the balloon moves toward the center of the curve (zero radius), rather than away from the center. The helium in the balloon, unaffected by the momentum acting on the air in the car, stays motionless relative to the ground outside the car.

Plunger Power

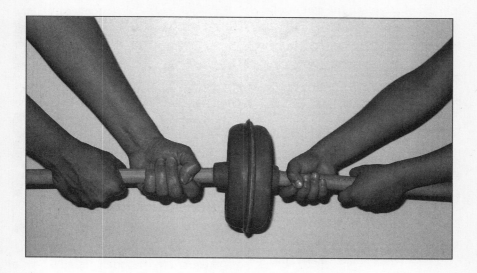

WHAT YOU NEED
☐ **Two plungers**
☐ **An assistant**

WHAT TO DO
Position the rims of the rubber cups on the ends of the plungers together and push firmly together until the suction cups collapse into each other.

Have an assistant grasp the wooden pole on one plunger while you grasp the second wooden pole. Try to pull the two plungers apart.

WHAT HAPPENS
The rubber plungers remain stuck together, no matter how hard you pull.

WHY IT WORKS
Pushing the two rubber cups together forces the air from between the cups and produces a vacuum in the enclosed space. Since no air can get inside the two cups, they remain vacuum-sealed together. The only way to separate the two plungers is to break the seal by prying the plungers apart and allowing air inside.

BIZARRE FACTS
▨ A plunger is a large suction cup.
▨ Trumpet players frequently use the rubber cup of a plunger to muffle the sound emitted from their horn.

- A plunger can be used to pull out a shallow dent in a car door. Simply wet the plunger, place over the dent, and pull out abruptly.
- Before drilling into a ceiling, cut a small hole through the center of the rubber cup of a plunger, and place the cup over the drill bit to catch falling chips of plaster.

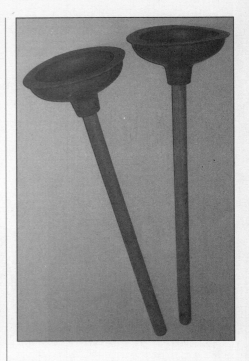

Take the Plunge

In the 1960s, a hippie in North Miami Beach, Florida, was seen carrying a plunger and wearing a burlap vest emblazoned on the back with the phrase: "Make Love, Don't Plunge into War."

- Insert the wooden handle of a plunger into the ground outdoors to create a candleholder.

Potato Radio

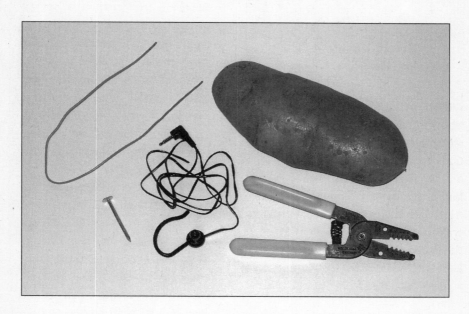

WHAT YOU NEED

- ☐ Wire cutters
- ☐ 12-inch-long 18-gauge copper wire
- ☐ Earphone from a transistor radio
- ☐ One galvanized nail (1 inch long)
- ☐ Potato

WHAT TO DO

With the wire cutters, cut the wire in half, and strip 1 inch of insulation off both ends of both pieces of wire. Wrap the end of one wire around the nail, just below the head. Insert the nail into one side of the potato.

Insert one stripped end of the second copper wire into the potato (without letting the wire and the nail touch each other inside the potato).

Cut off the plug at the end of the earphone wire, separate the two wires for 6 inches, and expose the two wires.

Attach one of the exposed wires from the earphone to one of the copper wires coming from the potato.

Place the earphone in your ear. Touch the second earphone wire to the free end of the second copper wire.

WHAT HAPPENS

You hear static through the earphone.

WHY IT WORKS

The citric acid in the potato acts as an electrolyte, conducting an electron flow between the copper in the wire and the iron in the nail, turning the potato into a battery. The resulting electrical current flows in the earphone wires, causing a static sound when the wires touch, when contact is broken, or when you rub the wires against each other.

BIZARRE FACTS

■ The leaves of the potato plant are poisonous if eaten.

■ The eyes in a potato are the indents where sprouts grow. You can grow potatoes by chopping up a potato into chunks with at least two or three eyes, letting the chunks dry in the sun for twenty-four hours, and then planting them.

■ Store-bought potatoes are frequently treated with a sprouting inhibitor. If you plant them, they may not grow.

■ Vincent Van Gogh's 1885 painting *The Potato-Eaters* portrays a family

Small Potatoes

Shortly after World War II, inventor George Lerner created a set of plastic noses, ears, eyes, and mouth parts that could be pushed into fruits or vegetables to create comical food characters. Toy companies initially rejected Lerner's new toy, however, convinced that American consumers, still clinging to a World War II mentality to conserve resources, would refuse to buy any toy that wasted food. But in 1952, Hasbro, Inc., a toy company based in Pawtucket, Rhode Island, launched Mr. Potato Head, a box of plastic parts (eyes, ears, noses, and mouths) that children could use to adorn a real potato (provided by Mom and Dad). One year after introducing Mr. Potato Head, Hasbro launched Mrs. Potato Head. On February 11, 1985, after twenty-three years of marriage, Hasbro introduced the Potato Heads' first tator tot—Baby Potato Head.

of five peasants gathered around a table, eating potatoes.

■ All Blue, Purple Peruvian, and Purple Viking are varieties of the potato with bright purple skin.

- In 1963, J. East of Spalding, Lincolnshire, Great Britain, grew the largest recorded potato in history, weighing seven pounds, one ounce. In 1982, J. Busby of Atherstone, Warwickshire, Great Britain, tied that record.
- In India, McDonald's offers the McAloo Tikki burger, a spicy vegetarian patty made of potatoes and peas.

Preserved Flowers

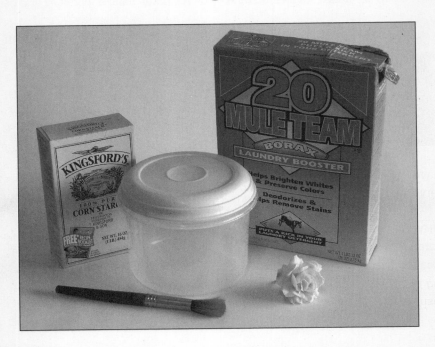

WHAT YOU NEED

- ☐ 20 Mule Team borax
- ☐ Cornstarch
- ☐ Airtight storage container with lid
- ☐ Flowers
- ☐ Soft artist's paintbrush

WHAT TO DO

Mix one part borax and two parts cornstarch. Fill the bottom inch of an empty airtight container with the mixture. Place a flower on the mixture, then gently cover the flower with more mixture, being careful not to crush the flower or distort the petals.

Flowers with a lot of overlapping petals, such as roses and carnations, are best treated by sprinkling the mixture directly into the blossom before placing them into the box. Repeat with more flowers. Seal the container and store at room temperature in a dry place for seven to ten days. When the flowers are dried, pour off the mixture and dust the flowers with a soft artist's brush.

WHAT HAPPENS

The borax and cornstarch remove the moisture from blossoms, petals, and

leaves, leaving you with perfectly preserved dry flowers.

WHY IT WORKS

Borax and cornstarch are hygroscopic, meaning they absorb water, preventing the wilting which would normally result as cut flowers age.

BIZARRE FACTS

- Fear of flowers is called anthophobia.
- In 1850, French novelist Alexander Dumas, author of *The Count of Monte Cristo*, *The Three Musketeers*, and *The Man in the Iron Mask*, published his novel *The Black Tulip*, a romantic tale that created popular fervor for the fictional black tulip.
- "Poppies! Poppies will make them sleep!" chants the Wicked Witch of the West in the 1939 movie classic *The Wizard of Oz*, as she casts a spell over a field of flowers. In the book, *The Wonderful Wizard of Oz*, written by L. Frank Baum in 1900, thousands of field mice pull the Cowardly Lion out of the Deadly Poppy Field. In the movie, Glinda—the Good Witch of the North—saves Dorothy, Toto, and the Lion by smothering the deadly scent of the poppies with snow.
- In Jamaica and other islands in the West Indies, the sap from the petals of hibiscus flowers is used as shoe polish.
- In Lewis Carroll's 1872 children's book *Through the Looking-Glass*, when Alice asks why she has never heard flowers talk in other gardens, the Tiger lily replies, "In most gardens they make the beds too soft—so that the flowers are always asleep."
- Carnations, daylillies, marigolds, nasturtiums, pansies, roses, and violets are all edible flowers.

Flower Power

Former Beatle George Harrison dedicated his 1980 book *I, Me, Mine* to "gardeners everywhere." In the book, he wrote, "I'm a gardener. I plant flowers and watch them grow. I don't go out to clubs and party. I stay at home and watch the river flow." The booklet to his 2002 posthumous album *Brainwashed* includes a photograph of his garden of topiaries.

Quaker Oats Planetarium

WHAT YOU NEED

- ☐ Black paint
- ☐ Paintbrush
- ☐ Clean, empty Quaker Oats canister with several plastic lids
- ☐ Large nail
- ☐ Flashlight
- ☐ Pencil
- ☐ X-acto knife
- ☐ Electrical tape

WHAT TO DO

Paint the inside of the lids with black paint. Let dry.

Using the nail, punch holes in the lids following the patterns of star for-mations or constellations, such as the Big Dipper, Orion, Pegasus, Taurus, or Leo. You can find the patterns in as-tronomy books or under "Astronomy" in an encyclopedia.

Hold the flashlight against the cen-ter of the bottom of the cardboard can-ister and draw a circle around it. With adult supervision cut out the circle with the X-acto knife. Secure the flash-light over the hole with strips of elec-trical tape. Wrap some electrical tape around the circumference of the head of the flashlight to prevent light from shining through the plastic or seeping through any cracks in the cardboard

other than the hole. Cover the canister with one of the hole-punched lids.

In a dark room, aim the lid of the canister at the ceiling or wall and turn on the flashlight. Gently turn the flashlight.

WHAT HAPPENS

The light box projects groups of small, starlike spots of light on the ceiling or wall, duplicating the constellations and creating a miniature planetarium.

WHY IT WORKS

When the light emitted from the flashlight travels through the small holes in the lid of the canister, it is projected in straight lines toward the ceiling or wall.

BIZARRE FACTS

- While the night sky seems to be filled with millions of stars, only some six thousand stars shine bright enough to be seen by the human eye without a telescope.
- Shooting stars—known as meteors—are actually chunks of rock or metal burning up as they enter the earth's atmosphere. Most meteors burn up before hitting the earth's surface. Those that do hit the ground are called meteorites.
- A person living at the equator can see all the stars visible from the earth during the year. A person at the North Pole cannot see the stars visible from the southern hemisphere. A person at the South Pole cannot see the stars visible from the northern hemisphere.
- Every night the stars appear to move from east to west because we on earth are moving from west to east. Only Polaris, better known as the North Star, does not appear to move because it is located almost directly above the North Pole. Since ancient times, navigators have used the North Star as a guide.
- As the earth orbits the sun over the period of a year, we see different constellations in the night sky that are otherwise obstructed from view by the light of the sun.
- The Big Dipper and Little Dipper are star groups called *asterisms* that are a part of two different constellations. The Big Dipper is a part of Ursa Major, and the Little Dipper is a part of Ursa Minor.
- To map the stars in the sky, astronomers divided the stars into eighty-eight constellations.
- As early as 2000 B.C.E., the Babylonians, convinced that the positions of stars and planets influence what happens on earth, mapped the posi-

tions of the stars and visible planets to predict events on earth. The ancient Egyptians, Greeks, and Romans also practiced astrology, but in the eighteenth century, most scientists rejected astrology as a false science, noting that events on earth and in space are based on the laws of physics and chemistry.

Ooops

Ancient Greek philosopher and scientist Aristotle (384–322 B.C.E.) incorrectly insisted that the earth was the center of the universe and that the sun rotated around the earth. These erroneous ideas were accepted as fact for nearly two thousand years and embraced by the Vatican, resulting in the persecution of several great scientific minds.

In 1592, the Inquisition arrested Italian scientist Giordano Bruno and jailed him for seven years as he stood trial for heresy for denying that the earth was flat and insisting that the universe was infinite, life existed on other planets, the sun did not revolve around the earth, and the earth was not the center of the universe. Eight years later, Pope Clement VIII sentenced Bruno to be burned alive at the stake.

In 1633, the Vatican summoned astronomer Galileo Galilei before the Inquisition and threatened to burn him at the stake for stating that the earth revolves around the sun in his 1632 book *Dialogue Concerning the Two Chief World Systems*. Instead, Galileo retracted his discoveries and spent the remaining eight years of his life under house arrest. A mere 359 years later, on October 30, 1992, Pope John Paul II, having conducted a thirteen-year investigation into the matter, formally stated that the Church was wrong to have condemned Galileo.

Rising Cardboard

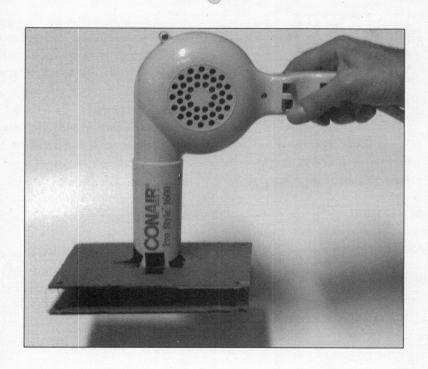

WHAT YOU NEED

- ☐ Blow-dryer
- ☐ Two squares of corrugated cardboard (7 by 7 inches)
- ☐ Pencil
- ☐ X-acto knife
- ☐ Hole puncher
- ☐ Dental floss
- ☐ Electrical tape

WHAT TO DO

Place the nozzle of the blow-dryer in the center of one of the squares of cardboard and trace around it with a pencil. With adult supervision, use the X-acto knife to cut out the disk.

Using the hole puncher, punch a hole in all four corners of both squares of cardboard.

Place the two pieces of cardboard together, and use the dental floss to make four loops through the four corners so that the two cardboard squares hang ½ inch apart.

Place the nozzle of the blower over the hole and tape it in place with the electrical tape.

Hold the blow-dryer so that the cards face downward and blow cool air.

WHAT HAPPENS
The bottom card rises up and presses against the top card.

WHY IT WORKS
The air moving between the two cardboard squares has less lateral air pressure than the still air that pushes up against the bottom of the lower card. This causes the bottom card to rise, in accord with Bernoulli's principle, first observed in 1738 by Swiss mathematician Daniel Bernoulli.

BIZARRE FACTS
▓ In nineteenth-century England, shirtmakers used hand-cranked corrugated roller presses to make ruffled shirt collars and cuffs. In 1856, shirt-makers adapted this press to create the world's first corrugated paper to make cylindrical liners to wrap around the inside of top hats to help them keep their shape. In 1871, American manufacturers first used corrugated paper to wrap bottles and glass chimneys for kerosene lamps. In 1894, paper manufacturers developed the first corrugated cardboard boxes.

▓ To make corrugated cardboard, a fluted layer is sandwiched between a bottom and top layer of linerboard. The flutes—essentially a series of connected arches—give the corrugated box extraordinary strength.

▓ The 1942 movie *Her Cardboard Lover,* directed by George Cukor, tells the story of a flirtatious lady who hires a lover to make her fiancé jealous.

The More Things Change

According to a 2001 report by the American Forest and Paper Association, most of the cardboard that is recycled is used to make more cardboard.

Rubber Band Ball

WHAT YOU NEED

☐ Bag of colored #64 rubber bands

WHAT TO DO

Tie a thick rubber band in a double knot. Wrap other bands around the knotted rubber band. The ball will look awkward until you achieve the size of a golf ball. Continue wrapping rubber bands around the ball until you achieve the size of a baseball. To maintain a spherical shape, place rubber bands over flat or bare spots. On a hard surface, bounce the ball.

WHAT HAPPENS

The ball will bounce like a rubber ball.

WHY IT WORKS

Rubber bands form the circumference of a circle, and a sphere is a circle spun around an axis. The rubber bands are made from rubber, and thus the resulting sphere is a rubber ball that bounces.

BIZARRE FACTS

■ Ancient Greek mathematician Euclid assumed that space was flat and, based on this assumption, wrongly concluded that parallel lines never meet. In the nineteenth century, German mathematician Georg Friedrich Bernhard Riemann proved that on the surface of a sphere, parallel lines do meet.

- In 1839, Charles Goodyear invented the vulcanization process for rubber after a long and courageous search that bordered on obsession. He died penniless, but today Goodyear tires are found on millions of cars.

- In 1876, Brazilian customs officials allowed English botanist Sir Henry Wickham to take seventy thousand rubber tree seeds from Brazil to England. Wickham insisted the seeds were botanical specimens to be used solely for the royal plant collection at Kew Gardens. Wickham germinated the seedlings at Kew Gardens, then sent them to Ceylon and Malaya, breaking the Brazilian monopoly on raw rubber.

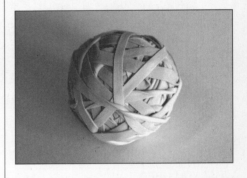

- In 1998, the *Guinness Book of World Records* acknowledged that John Bain of Wilmington, Delaware, single-handedly made the world's largest rubber band ball while working in the mail room of a law firm. The ball weighed 2,008 pounds with a circumference of 13 feet, 8½ inches, and comprised 350,000 rubber bands. You can see pictures of Bain's rubber band ball on his website: www.recordball.com.

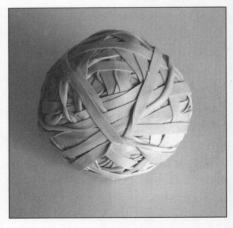

Rubber Bands from Heaven

In March 2003, Tony Evans of Swansea, Wales, dropped his one-ton rubber band ball out of a plane over Arizona to see if it would bounce. The television show *Ripley's Believe It or Not!* paid for the ball to be dropped out of the plane above the Mojave Desert and a skydiving cameraman filmed its descent. The ball took twenty seconds to hit the ground and created a four-foot-wide crater. It did not bounce. Instead, the ball broke apart on impact, leaving the rubber band remains in the bottom of the crater.

Steamboat

WHAT YOU NEED

- ☐ Clean, empty 1-liter plastic soda bottle
- ☐ X-acto knife or single-edge razor blade
- ☐ Hole puncher
- ☐ Sandpaper
- ☐ Copper tubing (2 feet long with ⅛-inch diameter)
- ☐ Pencil
- ☐ Electrical tape
- ☐ Votive candle
- ☐ Matches

WHAT TO DO

With adult supervision, cut the bottle in half lengthwise. Using the hole puncher, punch two holes 2 inches apart in the bottom of the bottle. Make sure the holes are parallel to the cut edge in the bottom of the bottle and equidistant from the center of the bottom.

Using sandpaper, sand the ends of the copper tube to prevent getting cut by sharp edges. Wrap the center of the copper tube around a pencil twice to make a coil.

Insert the ends of the tube through the two holes in the bottom of the bottle so that the coil rests inside the bottle, just below the neck. Tape the coil to the bottle so it stays in place. Tape the votive candle to the inside of the bottle under the coil of tubing and bend the coil to position it above the wick.

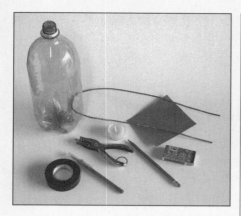

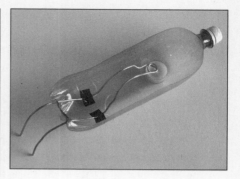

Set the boat in a bathtub filled part-way with water. Tilt the boat so one end of the tube is submerged in water and the other end is in the air. Suck on the free end of the tube until the tube is filled with water, then place the boat back in the water so both tubes are submerged.

Light the candle and observe.

WHAT HAPPENS

The boat travels the entire length of the bathtub, chugging along.

WHY IT WORKS

Heat from the candle flame converts water in the tube into steam. The steam (expanded water molecules that fill more space than the water) forces the remaining water out of the tube and propels the boat forward—in keeping with Sir Isaac Newton's third law of motion that for every action there is an equal and opposite reaction. The steam quickly cools, condenses, and water refills the tube. As long as the candle continues to burn, this process repeats itself, propelling the boat forward with a series of pulses.

BIZARRE FACTS

■ In 1698, Thomas Savery patented the first practical steam engine, a pump to drain flooded mines in Cornwall, England. In 1712, English blacksmith Thomas Newcomen invented a steam engine that powered a piston. In 1763, Scottish engineer James Watt improved upon a Newcomen engine by introducing the idea of two separate chambers for vaporization and condensation, rather than one chamber, which wasted enormous amounts of steam. He received a patent for his innovation in 1769.

■ In 1770, French army captain Nicolas Joseph Cugnot successfully operated a three-wheeled steam-powered automobile.

■ In 1928, Walt Disney titled his first Mickey Mouse cartoon *Steamboat Willie*.

■ In his song "Steam," musician Peter Gabriel sings the praises of Bubble Wrap.

- Steam shovels and steamrollers are powered by steam.
- In his song "Steamroller," James Taylor claims to be a steamroller and promises to roll over the woman he loves.

Letting Off Steam

Contrary to popular belief, Robert Fulton did not invent the steamboat. French engineer Denis Papip may have run one of the first steamboats on the River Fulda in 1707. In the United States, John Fitch began operating his steamboat on the Delaware River in 1787 and by 1790 ran regularly scheduled trips between Philadelphia and Burlington, New Jersey. Fulton invented his model of the steamboat in 1807, and his skill at marketing the safety and usefulness of the steamboat prompted people to incorrectly assume he had invented it.

Submarine

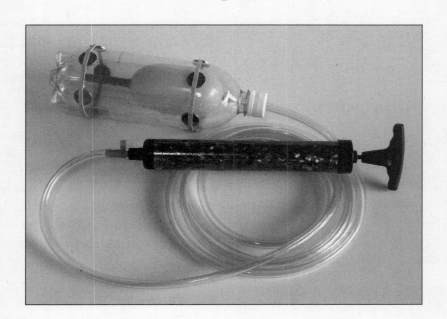

WHAT YOU NEED

- ☐ Clean, empty 1-liter plastic soda bottle with cap
- ☐ X-acto knife
- ☐ Three rubber bands
- ☐ Ten nickels
- ☐ Electric drill with ¼-inch bit
- ☐ 6-foot-long flexible plastic tube (¼-inch diameter)
- ☐ Balloon

WHAT TO DO

With adult supervision, cut five square 1-inch windows along the length of the bottle.

Put one rubber band around the top of the bottle and a second rubber band around the bottom of the bottle. Put five nickels under each rubber band, equidistantly surrounding the bottle.

Drill a ¼-inch hole in the middle of the cap. Fit one end of the tubing through the cap so that it extends 4 inches through the inside of the cap.

Blow up the balloon and then let the air out, so you have stretched the rubber. Put the mouth of the balloon around the end of the 4-inch length of

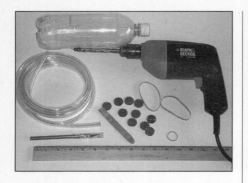

WHAT HAPPENS

When you put the submarine in the water, the water enters the holes, filling the bottle with water and causing the bottle to sink. To make the submarine surface, blow air into the tube. The air inflates the balloon, which expands and pushes the water out of the bottle, causing the submarine to float.

WHY IT WORKS

The water-filled submarine is negatively buoyant—or denser than the water (due to the additional weight of the nickels)—and the vessel sinks in the water. When the submarine is filled with air, it becomes buoyant—or less dense than the water—and floats to the surface.

BIZARRE FACTS

■ A submarine dives by achieving neutral or negative buoyancy by using ballast tanks that can be filled with either water, air, or a combination of both to adjust the ship's buoyancy. To make a submarine submerge, the crew lets water into the ballast tanks to make the ship heavier. To make the submarine rise, the crew uses compressed air to push the water out of the tanks, reducing the ship's weight. A propeller moves the submarine forward.

■ In 1620, the first known submarine, the *Drebbel*, was built for King James I of England. It navigated the Thames

tube and secure it in place with a rubber band.

Insert the balloon on the end of the tube inside the bottle, and screw on the cap securely.

Holding the free end of the tube, place the submarine in a swimming pool or a bathtub filled with water.

River and moved by use of oars that extended through sealed oarlocks.

- In his 1870 science-fiction novel *Twenty Thousand Leagues under the Sea,* French author Jules Verne tells the tale of Captain Nemo, a mad sea captain who cruises beneath the oceans in his submarine, the *Nautilus.*

- During World War II, forty thousand Germans served aboard Nazi submarines. Only ten thousand survived.

- On the popular television series *Voyage to the Bottom of the Sea,* Commander Lee Crane, played by actor David Hedison, captained the *Seaview,* a glass-nosed atomic submarine, through weekly adventures. The show was based on the 1961 movie of the same name, starring Walter Pidgeon, Joan Fontaine, Barbara Eden, Peter Loree, and Frankie Avalon.

- In the 1968 animated movie *Yellow Submarine,* the Beatles travel aboard a yellow submarine through the Sea of Holes and the Sea of Time to save the citizens of Pepperland from the Blue Meanies.

- The 1981 movie *Das Boot* chronicles the claustrophobic conditions aboard a German submarine during World War II.

Dive! Dive! Dive!

In 1978, a worker dropped a fifty-cent paint scraper into a torpedo launcher of the United States nuclear submarine *Swordfish*—costing the United States government 171,000 dollars in repairs.

Traveling Salt

WHAT YOU NEED

☐ Drinking glass

☐ Water

☐ Salt

WHAT TO DO

Fill a drinking glass with water and add salt until no more will dissolve. Let the glass sit undisturbed for several weeks.

WHAT HAPPENS

The water will be gone from the glass and the salt will coat the inside of the glass and carry over to the outside of the glass.

WHY IT WORKS

The water evaporates, leaving a fine coat of salt inside the glass. Capillary action carries more salt solution up the inside wall of the glass and eventually to the outside of the glass, where it evaporates, leaving the salt.

BIZARRE FACTS

■ Capillary action—the movement of water within the spaces of a porous material—occurs because water molecules stick to each other and to substances like glass, cloth, organic tissues, and soil. Water climbs up a thin glass tube because the hydro-

gen in the water bonds with the oxygens (typically bonded to hydrogen) at the surface of the glass—until the pull of gravity is too strong for the liquid to overcome. The narrower the glass, the higher the water will climb, because a narrow column of liquid weighs less than a wide one.

■ The Dead Sea is neither dead nor a sea. As an inland body of water, the Dead Sea is actually a lake, the saltiest body of water in the world (seven to eight times saltier than the ocean), and the lowest point on earth at 1,302 feet below sea level. Brine shrimp and a few salt-tolerant microorganisms (*Halobacterium halobium* and *Dunaliella*) live in its waters.

■ The first written reference to salt is found in the story of Lot's wife, who was turned into a pillar of salt when she disobeyed the angels and looked back at Sodom.

■ The expression "He is not worth his salt" originated in ancient Greece where salt was traded for slaves.

■ Roman soldiers were paid "salt money"—*salarium agentum*, the origin of the English word *salary*.

■ The superstition that spilling salt brings bad luck may have its origins in Leonardo da Vinci's *The Last Supper*, which depicts an overturned salt cellar in front of Judas Iscariot. The French believed that throwing a pinch of salt over the shoulder would hit the devil in the eye, preventing any further foul play.

■ Morton Salt is named for company founders Joy and Mark Morton. An advertising agency developed the fa-

Salty Sea

The Caspian Sea, the largest inland body of water in the world, is actually a salt lake. A sea is a body of water connected to an ocean, like the Caribbean Sea, the Mediterranean Sea, and the Arabian Sea. The Caspian Sea, a body of water nearly the size of California, is surrounded by land.

mous umbrella girl trademark, depicting a little girl standing in the rain with an umbrella over her head and holding a package of salt tilted backward with the spout open and the salt running out. Joy Morton's son, Sterling Morton II, then president of the company, suggested the slogan "When It Rains It Pours" to convey the message that Morton salt would run even in damp weather.

■ Salt inhibits the growth of bacteria, yeast, and molds, working as a natural preservative in butter, margarine, salad dressings, sausages, cured meats, and various pickled products. Salt also plays a key role in the leavening of bread, the development of the texture and rind of natural cheeses, the bleached color of sauerkraut, and the tenderness of vegetables.

Twirling Lines

WHAT YOU NEED

- ☐ **Compass with pencil**
- ☐ **White posterboard**
- ☐ **Scissors**
- ☐ **Large nail**
- ☐ **Ruler**
- ☐ **Heavy black marker**
- ☐ **Turntable**

WHAT TO DO

Using a compass with a pencil, make a circle 12 inches in diameter on the piece of posterboard. Using the scissors, cut out the disk.

Use the nail to punch a hole in the center of the disk.

Use the ruler and the black marker to make a thick black line along the diameter of the circle. Make three more heavy black lines along the diameter of the circle as if marking the posterboard into eight equal pieces like a pizza.

Place the posterboard disk in the center of the turntable, and spin the turntable rapidly.

WHAT HAPPENS

As the posterboard disk spins, the straight lines appear curved.

WHY IT WORKS

The brain, unable to properly interpret the rotating straight lines, simplifies the complex stimulus and perceives the rotating lines as circles.

BIZARRE FACTS

■ The lyrics to the pop song "Spinning Wheels," performed by Blood, Sweat, and Tears, tells of the spinning wheels of a merry-go-round.

■ If you stare at the tires of a car as the car moves forward, the wheels appear to spin backward.

■ A person wearing a suit with vertical stripes appears thinner than that same person wearing a suit with horizontal stripes does.

House of Illusions

A house painted white appears larger than a house painted a dark color.

Upside-Down Water

WHAT YOU NEED
☐ Drinking glass
☐ Water
☐ Cardboard (4 by 4 inches)

WHAT TO DO
Fill the drinking glass to the brim with water. Gently place the piece of cardboard over the mouth of the drinking glass. Holding the cardboard in place, turn the drinking glass upside down. Release the piece of cardboard.

WHAT HAPPENS
The cardboard sticks to the rim of the glass and holds the water in the glass.

WHY IT WORKS
Atmospheric pressure—the force exerted by the weight of the air—holds the cardboard in place.

BIZARRE FACTS
■ Atmospheric pressure is the force exerted by the weight of air molecules. At sea level, the earth's atmosphere presses against you with a force of 14.7 pounds per square inch. The force exerted on one square foot is more than one ton. But we're not squished by this because the molecules of our own bodies exert an equal and opposite force.

■ Your ears pop and you need to breathe more rapidly on top of a tall mountain because the atmospheric pressure is less than at sea level. In other words, there are less air molecules (causing you to breathe faster to fill your lungs with oxygen). At high altitudes, the air molecules also weigh less (causing your ears to pop to balance the pressure between the outside and inside of your ears).

■ At sea level the density of air is approximately eight ounces per square foot. Both pressure and density decrease by about a factor of ten for every ten-mile increase in altitude.

■ British civil engineer Sir Thomas Bouch designed a two-mile-long bridge to cross the Tay River in Scotland from Newport to Dundee. The bridge was opened in 1878, and Bouch was knighted in June 1879. Unfortunately, Bouch did not make any provisions for the bridge to withstand the effects of wind pressure nor did he provide any continuous lateral wind bracing below the deck. On December 28, 1879, a hurricane hit the bridge, blowing thirteen spans of wrought-iron lattice girders into the river, bringing the Edinburgh mail train with them, killing all seventy passengers aboard.

■ Camels do not carry water in their humps. The hump is a food reserve made primarily of fat. The concentration of fat in the hump lets the camel lose heat freely from the rest of its body, conserving water. A camel can go for days or even months without water because, unlike other animals, camels retain urea and do not start sweating until their body temperatures reach 115 degrees Fahrenheit.

■ No ocean water flows into the Panama Canal. The water in the Panama Canal is freshwater, flowing from streams and lakes into Gatún Lake, formed by a dam on the Chagres River. Freshwater flows out of the canal into the Atlantic and Pacific Oceans.

■ Some fish can breathe out of water. Lungfish breathe with gills and a lunglike organ, enabling them to breathe air. The South American and African lungfish sleep out of water for months at a time buried in the mud of dried-up riverbeds. The mudskipper, found from Africa to the East Indies to Japan, uses its muscular fins to hop around out of water on mudflats. The walking catfish of tropical Asia, breathing with lunglike organs, can journey overland for several days by using its strong front fins and thrashing its tail.

■ The Channel Tunnel, nicknamed the Chunnel, between Cheriton, England, and Fréthum, France, is 31 miles long, with 24 of those miles underwater. The Seiken Rail Tunnel between the Japanese islands of Honshu and Hokkaido is 33.5 miles long, with 14.5 of those miles underwater, making it the longest railroad tunnel in the world. The Chunnel, however, is the longest underwater railroad tunnel in the world.

■ Icebergs are not frozen ocean water. Icebergs are huge masses of ice that break off from glaciers or ice sheets. Icebergs are made from snow that has been compressed into ice over thousands of years by its own weight. Snow melts into freshwater—as do icebergs.

Water Colors

The Black Sea, Red Sea, White Sea, and Yellow Sea are named after the color of their water. A layer with a high concentration of hydrogen sulfide makes the water in the Black Sea look dark. In the summer, a type of algae forms a reddish brown scum on the surface of the Red Sea. Frozen from September to June, the White Sea is usually white. Yellow mud from the Huang River washes into the Yellow Sea, giving the water along the shore a muddy yellow color.

Weight-Lifting Magic

WHAT YOU NEED

- ☐ Electric drill with $\frac{1}{16}$-inch bit
- ☐ Two 12-inch lengths of wood ($1\frac{1}{2}$ by $1\frac{1}{2}$ inches)
- ☐ Six eyelet screws
- ☐ One screw hook
- ☐ Two 5-foot lengths of clothesline rope
- ☐ Well-supported rafter or beam
- ☐ A bucket of water

WHAT TO DO

With adult supervision, drill two holes spaced 9 inches apart on one side of a block of wood. In the opposite side, drill two holes spaced 6 inches apart. Screw four eyelet screws into the holes.

Drill one hole in the middle of one side of the second block of wood. In the opposite side, drill two holes spaced 6 inches apart. Screw a hook into the single hole and screw two eyelet screws into the holes on the opposite side.

Thread the first length of rope through the two eyelets that are 9 inches apart on one side of the first block of wood and hang the block from the rafter.

Thread the second length of rope through the rings of the two wooden blocks in a zigzag pattern (just like in the photograph). Rest the bucket handle in the hook. Pull the free end of the rope.

WHAT HAPPENS

Pulling the rope through the eyelets enables you to lift the water-filled bucket with less effort than trying to pick up the bucket on your own.

WHY IT WORKS

The eyelets act like pulleys, increasing the force applied and requiring less effort—creating a mechanical advantage.

BIZARRE FACTS

- A pulley gives a mechanical advantage of two, meaning that the force required to lift the object by pulling on the free end of the rope is only half the weight of the object.
- A single pulley does not reduce the amount of force needed to lift the object, but a single pulley does make lifting easier by changing the direction of force required.
- When used with ropes or cables, a number of pulleys can be combined in a variety of arrangements to minimize the force needed to lift heavy loads.
- In German, a pulley is called a *Flaschenzug.*

Pulley Power

Greek mathematician Archimedes, considered the greatest mathematician of ancient times, invented the compound pulley to defend his native city of Syracuse, Sicily, against the Romans in the Second Punic War.

Yogurt Cup Anemometer

WHAT YOU NEED

- 2-foot length of wood (1½ by 1½ inches)
- 1-foot square of wood (¾ inch thick)
- Ruler
- Pencil
- Electric drill with Phillips head screwdriver bit and ¼-inch bit
- 1¼-inch drywall screw
- Scissors
- Drinking straw
- 1½-inch metal washer
- Elmer's Carpenter's Wood Glue
- Two 18-inch lengths of wood (¾ by ¾ inch)
- Hacksaw
- Hammer
- Chisel
- 2-inch screw
- Vegetable oil
- Four clean, empty yogurt cups
- Red acrylic paint
- Paintbrush
- Four thumbtacks
- Stopwatch

WHAT TO DO

With adult supervision, stand the 2-foot length of wood on one end and center the 1-by-1-foot block of wood on it. Use the drill with a Phillips head screwdriver bit to screw a drywall nail through the center of the block into the end of the piece of wood to make a stand.

Position the stand upright on its base and mark a dot in the center of the top of the wood post. On the dot, drill a ¼-inch-diameter hole 2 inches deep into the stand.

Using scissors, cut a 2-inch length from the drinking straw, and fit the piece inside the drilled hole until the end of the straw is level with the opening of the hole. Glue the washer to the top of the wood so the hole in the washer lines up with the hole in the wood.

Make a cross with the two 18-inch lengths of ¾-by-¾-inch wood. In the center of the cross, use the saw and chisel to cut out a ⅜-inch-deep notch in each crosspiece. Glue the cross together and let dry.

Drill a ¼-inch hole through the center of the cross, and place a 2-inch-long screw through the hole.

Put a few drops of vegetable oil on the washer to reduce friction. Rest the cross horizontally over the hole in the top of the stand so the end of the screw fits inside the hole.

Paint the outside of one of the yogurt cups with red acrylic paint. Let dry.

Stick a thumbtack through the inside center of each yogurt cup and tack each yogurt cup horizontally to a different end of the cross, with the open ends of the cups facing the same direction. Use a few drops of wood glue between the cup and the wood to strengthen the bond. Let dry.

Place the anemometer outside in the wind. Using a stopwatch, count the number of times the red yogurt cup circles around in one minute.

Multiply the result by 4.71.

WHAT HAPPENS

The result is the wind speed in feet per minute. To convert the wind speed to miles per hour, multiply your result by .0114.

WHY IT WORKS

The cups catch the wind, which turns the cross. The circumference of the anemometer (the distance traveled by the red yogurt cup around the circle) equals the diameter of the circle (1.5 feet) multiplied by pi (3.14).

BIZARRE FACTS

- Ancient Greek physician Hippocrates, considered the father of modern medicine, claimed that south winds cause deafness and north winds cause constipation.
- The speed of wind is measured in miles per hour or knots (nautical miles per hour).
- On November 7, 1940, four months after the ribbon was cut on the $6.5 million Tacoma Narrows Bridge in Washington, a forty-two-mile-per-hour wind created oscillations in the suspension bridge (then the third longest in the world), tearing several suspenders loose and causing the single 2,800-foot span to break apart and fall into Puget Sound. Engineers had designed "Galloping Gertie" without taking into account aerodynamic effects and other structural weaknesses.

- In 1974, ABC-TV refused to allow the character Fonzie on *Happy Days* to wear a black leather jacket because it made him look like a hoodlum. In the first few episodes, Fonzie (played by Henry Winkler) wears a light-gray windbreaker. Producer Garry Marshall convinced the network to let Fonzie wear a black leather jacket and boots whenever he was near his motorcycle. Marshall then instructed his staff to make sure Fonzie was with his motorcycle in every scene. Fonzie became one of the most popular characters on television, and in 1980, Fonzie's black leather jacket was enshrined in the Smithsonian Institute.
- In the 1984 movie *This Is Spinal Tap,* directed by Rob Reiner, the fictional heavy metal band Spinal Tap releases an album called "Break Like the Wind."
- The 605-foot-tall Space Needle in Seattle, Washington, fastened to its

Gone with the Wind

In the 1939 movie *Gone with the Wind,* Melanie's pregnancy, when calculated by the dates of the Civil War battles mentioned, lasts twenty-one months.

foundation with seventy-two bolts, each measuring thirty feet in length, sways approximately one inch for every ten miles per hour of wind. Engineers built the Space Needle to withstand a wind speed of two hundred miles per hour.

Bibliography

The Book of Lists by David Wallechinsky, Irving Wallace, and Amy Wallace (New York: Bantam, 1977)

Dictionary of Trade Name Origins by Adrian Room (London, Routledge & Kegan Paul, 1982)

Duct Tape Book Two—Real Stories by Jim and Tim (Duluth, Minnesota: Pfeifer-Hamilton, 1995)

Einstein's Science Parties by Shar Levine and Allison Grafton (New York: John Wiley & Sons, 1994)

Elements of Psychology: A Briefer Course by David Krech, Richard S. Crutchfield, and Norman Livson (New York: Knopf, 1970)

Famous American Trademarks by Arnold B. Barach (Washington, D.C.: Public Affairs Press, 1971)

The Guinness Book of Records, edited by Peter Matthews (New York: Bantam, 1993, 1998)

How to Spit Nickels by Jack Mingo (New York: Contemporary Books, 1993)

Janice VanCleave's 200 Gooey, Slippery, Slimy, Weird & Fun Experiments by Janice VanCleave (New York: John Wiley & Sons, 1993)

The Joy of Cooking by Irma S. Rombauer and Marion Rombauer Becker (New York: Bobbs-Merrill, 1975)

Jr. Boom Academy by B. K. Hixson and M. S. Kralik (Salt Lake City, Utah: Wild Goose, 1992)

Martin Gardner's Science Tricks by Martin Gardner (New York: Sterling, 1998)

Modern Chemical Magic by John D. Lippy, Jr., and Edward L. Palder (Harrisburg, Pennsylvania: Stackpole, 1959)

More Science for You: 112 Illustrated Experiments by Bob Brown (Blue Ridge Summit, Pennsylvania: TAB Books, 1988)

100 Make-It-Yourself Science Fair Projects by Glen Vecchione (New York: Sterling, 1995)

PADI Open Water Diver Manual (Rancho Santa Margarita, California: International PADI, 1999)

Panati's Extraordinary Origins of Everyday Things by Charles Panati (New York: Harper & Row, 1987)

Reader's Digest Book of Facts, edited by Edmund H. Harvey, Jr. (Pleasantville, New York: Reader's Digest, 1987)

Reader's Digest How Science Works by Judith Hann (Pleasantville, New York: Reader's Digest, 1991)

Ripley's Believe It or Not! Encyclopedia of the Bizarre by Julie Mooney and the Editors of *Ripley's Believe It or Not!* (New York: Black Dog & Leventhal, 2002)

Science Fair Survival Techniques (Salt Lake City, Utah: Wild Goose, 1997)

Science for Fun Experiments by Gary Gibson (Brookfield, Connecticut: Copper Beech Books, 1996)

Science Wizardry for Kids by Margaret Kenda and Phyllis S. Williams (Hauppauge, New York: Barron's, 1992)

Shout!—The Beatles in Their Generation by Philip Norman (New York: Warner Books, 1981)

Sure-to-Win Science Fair Projects by Joe Rhatigan with Heather Smith (New York: Lark Books, 2001)

333 Science Tricks & Experiments by Robert J. Brown (Blue Ridge Summit, Pennsylvania: TAB Books, 1984)

365 Simple Science Experiments by E. Richard Churchill, Louis V. Loeschnig, and Muriel Mandell (New York: Black Dog & Leventhal, 1997)

200 Illustrated Science Experiments for Children by Robert J. Brown (Blue Ridge Summit, Pennsylvania: TAB Books, 1987)

The Ultimate Duct Tape Book by Jim and Tim (Duluth, Minnesota: Pfeifer-Hamilton, 1998)

The Way Science Works by Robin Kerrod and Dr. Sharon Ann Holgate (New York: DK Publishing, 2002)

Why Did They Name It . . . ? by Hannah Campbell (New York: Fleet, 1964)

Acknowledgments

I am deeply grateful to my editor, Michelle Howry, for her enthusiasm, keen insights, and professionalism; John Duff at Perigee Books for his encouragement, and sharp wit; Sheila Moody for her astute copyediting skills; and my wife Debbie for her outstanding research.

A very special thanks to my agent, Jeremy Solomon, for making this book possible.

I am also grateful to Haleigh Safran, Sue Solomon, and Matt Strauss at *The View;* Becky Cohen at *The Other Half;* Mark, Debbie, Marissa, and Jason Jaffe for providing the inspiration for Mayonnaise Madness; and Jim Parish for suggesting the idea in the first place.

Above all, all my love to Debbie, Ashley, and Julia, for being my able assistants in the garage, and for always giving me the honor and pleasure of cleaning up the mess.

About the Author

Mad Scientist Joey Green got Barbara Walters to make green slime and showered Meredith Vieira with a geyser of exploding Coca-Cola on *The View*. Together with Dick Clark, he put actor Mario Lopez, star of *Saved by the Bell*, inside a giant monster bubble on *The Other Half*. Joey has also gotten Jay Leno to shave with peanut butter on *The Tonight Show*, Rosie O'Donnell to mousse her hair with Jell-O on *The Rosie O'Donnell Show*, and Katie Couric to drop her diamond engagement ring into a glass of Efferdent on *Today*. He has been seen polishing furniture with Spam on *Dateline NBC*, cleaning a toilet with Coca-Cola in the pages of the *New York Times*, and washing his hair with Reddi-wip in *People*. Green, a former contributing editor to *National Lampoon* and a former advertising copywriter at J. Walter Thompson, is the author of thirty books, including *Polish Your Furniture with Panty Hose*, *Clean Your Clothes with Cheez Whiz*, *The Zen of Oz*, and *The Road to Success Is Paved with Failure*—to name just a few. A native of Miami, Florida, and a graduate of Cornell University, he wrote television commercials for Burger King and Walt Disney World, and won a Clio Award for a print ad he created for Eastman Kodak. He backpacked around the world for two years on his honeymoon, and lives in Los Angeles with his wife, Debbie, and their two daughters, Ashley and Julia.

Visit Joey Green on the Internet at www.wackyuses.com.